Leonard Waldo

Report of the observations of the total solar eclipse, July 29, 1878

Leonard Waldo

Report of the observations of the total solar eclipse, July 29, 1878

ISBN/EAN: 9783743376922

Manufactured in Europe, USA, Canada, Australia, Japa

Cover: Foto ©berggeist007 / pixelio.de

Manufactured and distributed by brebook publishing software
(www.brebook.com)

Leonard Waldo

Report of the observations of the total solar eclipse, July 29, 1878

REPORT OF THE OBSERVATIONS

OF

THE TOTAL SOLAR ECLIPSE,

JULY 29, 1878,

MADE AT FORT WORTH, TEXAS.

EDITED BY

LEONARD WALDO,

ASSISTANT AT THE OBSERVATORY OF HARVARD COLLEGE.

CAMBRIDGE:

PRESS OF JOHN WILSON AND SON.

1879.

ACKNOWLEDGMENT.

———————◆———————

It would be ungracious to allow this monograph to find its way into the scientific world without recognizing the generosity of those gentlemen who have by their material aid assisted us in bringing to a gratifying conclusion our observations of the Total Solar Eclipse of 1878. We feel especially indebted to —

THE BOARD OF DIRECTORS OF THE BACHE FUND.
THE OFFICERS OF THE WESTERN UNION TELEGRAPH COMPANY.
THE OFFICERS OF THE PENNSYLVANIA RAILROAD AND OF THE RAILROADS CONNECTING ST. LOUIS WITH FORT WORTH.
THE OFFICERS OF WASHINGTON UNIVERSITY, ST. LOUIS.
THE OFFICERS OF THE UNIVERSITY CLUB, ST. LOUIS.
S. W. LOMAX, ESQ., AND THOSE OTHER CITIZENS OF TEXAS WHO ACTIVELY CO-OPERATED WITH US.

To the following gentlemen of St. Louis: —

GERARD B. ALLEN.
THOMAS ALLEN.
ROBERT B. BROWN.
C. B. BURNHAM.
GEORGE O. CARPENTER, JR.
DANIEL CATLIN.
M. DWIGHT COLLIER.
JOHN T. DAVIS.
W. G. ELIOT.
C. S. GREELEY.

W. H. GREGG.
EDWIN HARRISON.
E. A. HITCHCOCK.
HENRY HITCHCOCK.
GEORGE E. LEIGHTON.
CHARLES PARSONS.
J. C. RICHARDSON.
R. M. SCRUGGS.
W. A. THORNBURGH.
JAMES E. YEATMAN.

And to —

PROFESSOR S. S. GREENE, of Brown University, Providence.
E. ELLERY ANDERSON, Esq., of New York City.
T. P. SMITH, Esq., of Waltham, Mass.
D. T. SMITH, Esq., of Salem, Mass.
THE FIRM OF ALVAN CLARK & SONS, Cambridgeport, Mass.

We trust that these gentlemen will find in the following pages an indication that their aid has resulted in some slight accession to our knowledge of the Sun.

THE EDITOR.

CAMBRIDGE, January, 1879.

Contents.

————◆————

CONTENTS.

INTRODUCTION.

————•———

THE " Fort Worth Eclipse Party " consisted of five observers who organized the party with the special end in view of recording such phenomena as might aid in establishing the correct theory regarding the corona which is seen surrounding the sun during the moments of a total solar eclipse.

The members were : —

>   Mr. LEONARD WALDO, of Harvard College Observatory.
>   Mr. R. W. WILLSON, of Harvard College.
>   PROFESSOR J. K. REES, of Washington University, St. Louis.
>   Mr. W. H. PULSIFER, of St. Louis.
>   Mr. F. E. SEAGRAVE, of Providence, R.I.

The general plan of work to be done was proposed by myself, and after discussion of it by the party and modification in some of its details, it was adopted.

According to this scheme, it was determined that the corona should be examined with the naked eye, the telescope, the spectroscope, the polariscope, and that an effort should be made to photograph the corona, and also to take such photographs as might indicate polarization, by placing between the two lenses of cameras double refracting prisms.

Further, that we should make such other observations regarding the duration of totality, the times of contacts, and such miscellaneous observations as should be possible without interfering with our corona work, and that, in furtherance of this plan, we should encourage the general observance of the eclipse by such persons as might be available in the country about our station.

The following instrumental equipment was provided : —

1. A 5-inch telescope of 73 inches focus, made by Alvan Clark & Sons, mounted upon a steady stand and provided with a battery of eye-pieces, giving 58, 92, 150, 260 diameters respectively. Loaned by T. P. Smith, Esq., of Waltham, Mass.

2. A 5-inch telescope of 81 inches focus, made by Alvan Clark & Sons, mounted upon a portable tripod stand, with a battery of eye-pieces, giving 61, 100, and 134 diameters respectively. Loaned by Alvan Clark & Sons, of Cambridgeport.

3. A 4-inch telescope of 58 inches focus, made by Alvan Clark & Sons, mounted upon a substantial tripod, and provided with a battery of eye-pieces, giving 35, 75, 155, and 310 diameters respectively. Property of Mr. Pulsifer.

4. A 3-inch telescope of 45 inches focus, made by Secretan of Paris, mounted upon a substantial stand. Loaned by E. Ellery Anderson, Esq., of New York City.

5. A 3-inch telescope of unknown make. Property of Mr. Seagrave.

6. A draw telescope of 1.35 inches aperture, provided with an erecting eye-piece, giving a magnifying power of 24.5 diameters. Property of Mr. Waldo.

7. A draw telescope of 2 inches aperture.

8. A spectroscope by Browning, having a dispersive power of ten 60° flint glass prisms. Property of Mr. Seagrave.

9. A direct vision spectroscope by Browning, with a train of ten 60° prisms. Property of Mr. Pulsifer.

10. A spectroscope by William Grunow, having two large prisms by Courvoisier of Paris, each side of whose base is 1.5 inches, and which each give a minimum deviation of the D ray of about 48° 42′, and a dispersion between the extreme red and the extreme violet of 0.0833.

The telescopes are of 1.52 inches aperture and 14.0 inches focus, with triple object glasses by Browning. The instrument is provided with micrometric apparatus, and an additional pair of telescopes of 7 inches focus. Property of Mr. Waldo.

11. A single prism spectroscope by Alvan Clark & Sons. Property of Mr. Pulsifer.

12. A portable transit instrument by Temple, of 1.7 inches aperture and 28 inches focal length. Called "Temple Transit." Loaned by D. T. Smith, Esq., of Salem.

13. A portable transit instrument by Troughton and Simms, of 1.6 inches aperture and 23 inches focal length. Called "Fort Worth Transit." Loaned by Professor Greene, of Brown University, Providence.

14. Sidereal Box Chronometer, "Victor Kullberg, 1178." Property of Mr. Seagrave.

15. Mean Time Box Chronometer, "William Bond & Sons, 1058."

16. Mean Time Pocket Chronometer, "Johnson, 1436.", Property of Mr. Waldo.

17. Sextant, "Stackpole & Brothers, 1707." With stand designed by Mr. Willson.

18. Large polariscope for photographing an artificial eclipse. Loaned by the Rev. E. C. Bolles, Salem:

19. Two large double refracting prisms. Loaned by Professor A. Litton, of Washington University, St. Louis.

20. Photographic outfit furnished by our photographer, Mr. A. Freeman, of Dallas, Texas, consisting of three cameras complete with all accessories for taking the photographs.

21. Telegraphic instruments.

22. Four stop watches. Loaned by the Auburndale Rotary Watch Company.

23. Miscellaneous apparatus, consisting of a barometer, thermometers, small spectroscopes, Arago polariscope, glass scales ruled by Professor W. A. Rogers, eye-pieces, sun-shades, &c.

The selection of Fort Worth as the observing station was based upon the following considerations:—

1. The chances of a clear day were as great as for any point along the line of totality, according to the valuable paper on this subject put at our and other observers' disposal, by the United States Signal Service.

2. The totality would occur with us some fifteen minutes after it had occurred in Colorado,

which would cause the record in Texas to have additional value in questions relating to any rapid change in the corona.

3. There would be, no doubt, plenty of observers who would go to Colorado on such a mission; but it was not so certain that any other well-equipped party would go to Texas in the summer season.

Mr. Willson accordingly left St. Louis, July 8th, and proceeded to Fort Worth to arrange for the rest of the party who were to join him later. Before leaving St. Louis, however, he and Professor Rees had established the "Temple Transit" upon the transit piece in the observatory lately erected by the Washington University at the corner of St. Charles Street and Eighteenth Street.

I arrived in St. Louis the morning of July 15th, and at once proceeded, with the aid of Professor Rees, to get ready for the exchange of longitude signals between Fort Worth and St. Louis. In the evening, however, after having completed the first set of transit observations, and upon repairing to the telegraph office, we received word from Mr. Willson that he had been unable to mount his transit instrument as yet; and he further advised our proceeding to join him at Fort Worth, as he was somewhat under the influence of the weather. Accordingly, after a telegraphic comparison of chronometers, we prepared to leave St. Louis on the night of the 16th, which we did, arriving at Fort Worth the afternoon of the 18th of July.

Upon presenting the letters we bore to S. W. Lomax, Esq., he immediately tendered us the use of his grounds, and all the facilities his home could offer. The site was extremely favorable, about half a mile from the city limits, commanding an extensive view in all directions. We very gladly therefore accepted Mr. Lomax's kind offer, and chose the sites for our instruments to the west of his house.

Counting from the north toward the south, the observers Pulsifer, Rees, Willson, Seagrave, and Waldo were in order, and about 40 feet apart. The house-top, which fortunately was flat, afforded an excellent place for the sketchers and for the person who was to call off the seconds remaining of totality. Upon digging to the depth of about 3 feet, we found limestone rock; and our transit pier, which consisted of four upright scantlings about 3 × 4 inches, firmly boxed together, rested upon this bed of rock. A gravel and broken stone mixture was then filled in around the scantlings, and the interior of the pier up to the level of the ground was tamped with the same mixture. A wooden block about fourteen inches square made the top of the pier, and a tight box resting on this top was made to set over the transit, and protect it in bad weather. Two posts set about three feet from the E and W pivots held the lamps for illumination of the field of view, without danger of warming the pivots. Within a few feet of this pier, a post was set into the ground to support an artificial horizon at a convenient height for an observer sitting in a chair to use the sextant.

We had caused to be reprinted from the excellent "Instructions for observing the Total Solar Eclipse," prepared by Professor William Harkness at the direction of the Superintendent of the United States Naval Observatory, a pamphlet of extracts, including Section I. entire and that part of Section V. included between the sub-heading "The Contacts" and the end of the section. Copies of this reprint were furnished to each of the observers who wished to co-operate with us. A class was at once organized among those willing to aid us about

Fort Worth, and was exercised in sketching the outlines of the previous eclipses as shown in the various reports, the time of making a single sketch being limited to two and a half minutes.

The State press exerted itself in spreading the suggestions made to inexperienced observers, and the reports given in the sequel show that our efforts to interest others in this work were fruitful.

Regular routine was at once established. Messrs. Willson and Rees took the meridian transit observations, with the sextant Professor Rees observed the sun daily for latitude, and I observed α Scorpii and Polaris for the same purpose at night. Mr. Seagrave had the meteorological observations to make, and a daily observation of the sun's disc; in the evening, he assisted me. The direct preparation for the eclipse, however, was the most arduous task. The spectroscopic work must all be rehearsed; and, since it was extremely warm in the sun, it became a trying thing to work more than a few minutes at a time over instruments which were too warm to handle.

Mr. Pulsifer joined the party four days preceding the eclipse; and we immediately chose our assistants from the helpers who had gathered at Fort Worth for this purpose. A heavy storm beginning the Friday before the eclipse, and lasting over Saturday and Sunday, so seriously interfered with our preparations that we were fearful Monday would find us not in readiness. Particularly was I fearful about the photographs. Mr. Alfred Freeman could not come to Fort Worth until the Wednesday previous to the eclipse. Thursday and Friday had been fully occupied in rigging up a dark room and testing the chemicals. The cameras were all in readiness, but we had had no chance to try their rates of motion with the improvised clock-works. Nor did there any chance occur until Monday afternoon, and then I found it was too late to make any thing like a good adjustment of its motion; and so all of our photographs are "moved pictures," and therefore they only indicate the position of the prominences without showing much detail.

A carefully prepared and detailed scheme dividing the work during the eclipse among the various observers was arranged. By this scheme, the appointment of the instruments named in the preceding list and the work to be accomplished was as follows : —

Mr. R. W. Willson, the general telescopic and polariscopic phenomena.

Instruments, — Nos. 3, 15, Arago polariscope, hand spectroscope.

Professor J. K. Rees, general telescopic phenomena during the partial phases, and the measurement of coronal lines during totality; the extension of the corona away from the sun as determined spectroscopically.

Instruments, — Nos. 1, 10, and a stop watch.

Mr. W. H. Pulsifer, spectroscopic observations of the contacts; Young's reversion layer; the position of the " 1474 " line, and a determination as to whether it joins the sun's disc, is broader at its base than its apex, &c.; the examination of the prominences for new lines, and the cusps for rays.

Instruments, — Nos. 3, 5, 9, 11, and the watch, " Jurgensen, 13548."

Mr. F. E. Seagrave, spectroscopic observation of the contacts; the discovery of any new prominence lines, particularly beyond F.; the micrometric measurement of the line " 1474," and the meteorological observations.

Instruments, — Nos. 2, 8, 14, and thermometer "Casella, 29563."

As for myself, I expected to be fully occupied with the photographs during totality, and with the general work always necessary during the partial phases.

Instruments, — Nos. 6, 16, 20, and a stop watch.

Mr. A. M. Britton of the City National Bank occupied a position on the house-top, from which the approach of the eclipse could be seen; and it was his duty to call out in a strong tone the number of seconds remaining of totality, each 15 seconds after its commencement, assuming the totality to last 2^m 30'. He also was to obtain a determination of the duration of the totality with his stop watch, "Ls. Audemars, 11855." The class of sketchers were seated by Mr. Britton, and before each one of them a plummet was suspended by a cord, which should serve as vertical reference lines in their drawings.

Eight sketchers furnished drawings of the totality; and four of the drawings have been selected for publication, and are given in Plate IV.

Professor Lockett's drawing, which is the most valuable of the naked-eye sketches, was made by a careful draughtsman, and is a capital representation of the corona as I saw it with the unaided eye. In preparing all of our plates for publication, the original drawings were photographed directly on the block. They seem to me to be extremely faithful copies of the originals. Professor Lockett's drawing better gives the effect of an oil painting he has since made, by viewing it from a distance of about ten feet.

Phase.	Observer.	Fort Worth Mean Time.	Remarks.	Page of this Report.
		h. m. s.		
I Contact.	L. Waldo.	3 11 18.7	Small telescope. See Report.	30
	R. W. Willson.	11 10.7		40
	J. K. Rees.	11 11.3	Estimated. Telescopic contact.	47
	W. H. Pulsifer.		Lost.	49
	F. E. Seagrave.	10 44.4	Disappearance of the C line. 10 prisms.	53
II Contact.	L. Waldo.		Not observed.	30
	R. W. Willson.	4 17 6.1	This is probably recorded 10' early, as Mr. Willson	40
	J. K. Rees.	17 19.4	had been carrying the chronometer beat for 2^m	47
	W. H. Pulsifer.	17 15.3	in his head. — L. W.	50
	F. E. Seagrave.			53
III Contact.	L. Waldo.		Not observed.	30
	R. W. Willson.	4 19 47.	See Report.	40
	J. K. Rees.		Not observed.	47
	W. H. Pulsifer.		Not observed.	50
	F. E. Seagrave.		Not observed.	53
IV Contact.	L. Waldo.	5 19 14.5	Small telescope. Observed through clouds.	30
	R. W. Willson.	19 18.3		40
	J. K. Rees.	19 28.5	Cloudy.	47
	W. H. Pulsifer.		} Too cloudy to be seen with the spectroscopes. {	50
	F. E. Seagrave.			53
Duration of Totality.	A. M. Britton.	2 28.75		11

For convenience of reference, I have gathered together the observations which include the element of time; and I give them in the accompanying abstracts, the first of which relates to the observations made at our own station. The observations made on the day of the eclipse will be found recorded in the reports following the Introduction.

The approximate geographical positions of the following observers are given on page 27, and the details of the observations are given on the pages indicated in the last column:—

Locality.	Observers.	I Contact.	II Contact.	III Contact.	IV Contact.	Duration of Totality.	Local Time used.	Page of this Report.
ALLEN, Collin Co., Texas.	H. A. Hill, J. M. Hobson, L. M. Rush.	h. m. s. 3 19 8	h. m. s. 4 23 54	h. m. s. 4 25 39	h. m. s. 5 20 54	m. s. 1 45	Houston. Slow of Washington signals 14 s sm.	58
DALLAS, Dallas Co., Texas.	James Giles, J. H. Breeze.	3 11 0	4 19 58	4 21 56	5 20 0		Dallas.	59
McKINNEY, Collin Co., Texas.	J. M. Pearson, John S. Moore, W. J. Finch.	3 10 30	4 13 32		5 14 30	1 27	See Report.	59
„ „	W. H. Chandler.	3 8	4 14		5 14 30	1 29¾		60
BREMOND, Texas.	W. Quarles.					1 56½		60

Three of the party left Fort Worth on the evening of the 29th of July, Mr. Pulsifer and myself desiring to arrive at St. Louis as soon as possible, in order to exchange longitude signals with Messrs. Willson and Rees, who remained in Fort Worth some days longer for that purpose. I should mention, before closing the descriptive part of the Introduction, that Messrs. Lomax and Britton, in connection with a few other gentlemen interested in the geographical position of Fort Worth, subscribed a fund for the construction of a monument in the Court House Square at Fort Worth, whose position should be made to depend upon the position of our transit pier by means of a survey executed by Mr. Terry, and which may be found on page 27. The inscription furnished to Mr. Lomax for this pier was,—

N. Latitude, 32° 45′ 19″
W. Longitude, 1h 21m 7s.57

FORT WORTH ECLIPSE PARTY,
July 29, 1878.

W. H. PULSIFER.   J. K. REES.
R. W. WILLSON.   F. E. SEAGRAVE.
LEONARD WALDO.

and it is here given, so that it may be easily identified at any future date.

CHRONOMETER ERRORS AT FORT WORTH.

THE sidereal chronometer " Victor Kullberg, 1178," was used as the standard time-piece of the party. At ordinary temperatures, it performs admirably with a rate of about 2ˢ.6 losing. The extremely high temperatures to which all the chronometers were subjected at our observing station proved, however, destructive to regularity in their rates.

The observations for time were made by Mr. Willson, assisted by Professor Rees. The "Brown University" transit instrument was used, with the exception of two sextant sets made July 16th and 19th by Mr. Willson.

This transit is an old one by Troughton and Simms, of the pattern figured in Bowditch's Practical Navigator. A new level, kindly loaned to us by Messrs. Buff and Berger of Boston, was adapted to the instrument; and, after the date of July 26th, the ordinary web reticule of 5 webs was replaced by one ruled on glass by Professor W. A. Rogers, having three groups of lines of 3, 5, and 3 lines respectively. The corrections to the first series of webs to reduce each one to the mean were, for Circle E,

I.	II.	III.	IV.	V.
− 49ˢ.27	− 23ˢ.98	− 0ˢ.63	+ 24ˢ.72	+ 49ˢ.10

The Equatorial distance between two consecutive lines of the glass reticule was 12ˢ.633 and between the nearest lines of two consecutive groups 25ˢ.266, without any sensible variation. The value of a division of the Buff and Berger level I found to be

$$1^d.00 = 1''.470 \text{ at } 79°.5 \text{ F.}$$

which is the mean of ten settings of the Harvard College Observatory Meridian Circle. The extreme heat, the annoyance caused by the numerous insects which fluttered into the observers' faces, or crept under their clothing, or got into positions such that a movement of the instrument would jam them under the pivots, or under the level in reversing it, coupled with the fact that our whole observatory was a temporary one, all conspired to render our observations both for time and latitude less accordant than we wished. Still, I think all our determinations are fully within the limit of accuracy imposed by a rigorous conception of the chief objects of our expedition. Rejecting those nights which are not fairly accordant, the following are the time observations used : —

TIME OBSERVATIONS AT FORT WORTH.

Mean Date. 1878.	Star's Name.	Lamp.	Observed Time. Kullberg, 1178.	Aa.	Bb.	Cc.	Error, Kullberg, 1178.	Mean Error.
			h. m. s.	s.	s.	s.	h. m. s.	h. m. s.
	τ Herculis.	W.	17 59 41.32	− .24	+ 2.14	− 1.73	+ 1 43 34.77	
	x Ophiuchi.	W.	18 35 30.16	+ .28	+ .99	− 1.21	+ 1 43 34.11	
	δ Herculis.	W.	18 40 42.82	− .02	+ 1.15	− 0.96	+ 1 43 34.45	
	μ Herculis.	E.	19 25 17.12	+ .19	+ 1.06	+ 1.35	+ 1 43 34.54	
July 20.37								+ 1 43 34.45
	γ Draconis.	W.	19 37 5.20	+ 12.97	− .76	− 1.47	+ 1 43 26.20	
	γ⁹ Sagittarii.	W.	19 41 56.20	− 25.95	− .23	− 1.06	+ 1 43 26.36	
	μ Sagittarii.	W.	19 51 21.70	− 20.76	− .21	− .98	+ 1 43 27.70	
	η Serpentis.	W.	19 58 45.66	− 14.27	− .41	− .92	+ 1 43 26.78	
	α Lyræ.	E.	20 16 15.02	+ 3.24	− .98	+ 1.18	+ 1 43 26.87	
July 22.43								+ 1 43 26.78
	τ Herculis.	W.	17 59 28.70	+ 2.35	+ .45	− .70	+ 1 43 23.70	
	α Scorpii.	W.	18 5 29.98	− 6.57	+ .21	− .53	+ 1 43 23.50	
	A Draconis.	W.	18 11 29.58	+ 11.10	+ .78	− 1.34	+ 1 43 23.69	
	x Ophiuchi.	W.	18 35 23.02	− 2.69	+ .29	− .49	+ 1 43 23.25	
	α¹ Herculis.	E.	18 52 33.82	+ 2.40	+ .19	+ .15	+ 1 43 23.73	
	α Ophiuchi.	E.	19 12 44.80	− 2.52	+ .10	+ .50	+ 1 43 23.12	
July 23.37								+ 1 43 23.50
	τ Herculis.	W.	17 59 24.17	+ 1.26	+ .40	− 1.84	+ 1 43 16.94	
	α Scorpii.	W.	18 5 19.61	− 3.42	+ .17	− 1.37	+ 1 43 15.41	
	η Herculis.	W.	18 22 3.14	+ 0.50	+ .04	− 1.62	+ 1 43 16.34	
	α¹ Herculis.	E.	18 52 24.86	− 1.15	+ .88	+ 1.30	+ 1 43 17.52	
	44 Ophiuchi.	E.	19 2 18.52	− 3.27	+ .30	+ 1.36	+ 1 43 16.63	
	α Ophiuchi.	E.	19 12 36.20	− 1.26	+ .30	+ 1.29	+ 1 43 16.87	
July 25.37								+ 1 43 16.62
	α¹ Herculis.	W.	18 52 18.00		+ .40		+ 1 43 9.15	
	ϕ Draconis.	W. & E.	19 37 0.08		+ .79		+ 1 43 9.40	
	μ Sagittarii.	E. & W.	19 49 40.60		+ .24		+ 1 43 8.93	
July 27.40								+ 1 43 9.16
	δ Draconis.	W.	20 55 23.08	+ 12.83	+ .80	+ 5.20	+ 1 43 7.49	
	δ Aquilæ.	W.	21 2 33.97	− 4.30	+ .33	+ 2.00	+ 1 43 7.52	
	ι¹ Cygni.	W.	21 9 39.80	+ 4.38	+ .56	+ 3.20	+ 1 43 7.07	
	ι¹ Cygni.	E.	21 9 46.76	+ 4.01	+ .42	− 3.20	+ 1 43 7.12	
	ν Aquilæ.	E.	21 23 42.70	− 3.36	+ .23	− 2.05	+ 1 43 6.26	
	α Aquilæ.	E.	21 28 05.47	− 3.52	+ .19	− 2.04	+ 1 43 7.53	
July 28.48								+ 1 43 7.16
	α Scorpii.	W.	18 5 20.38	− 19.82	.00	+ .14	+ 1 43 1.17	
	η Herculis.	E.	18 21 43.46	+ 2.92	.00	− .15	+ 1 43 0.67	
	x Ophiuchi.	W.	18 35 5.44	− 8.08	+ .13	+ .12	+ 1 43 0.81	
	d Herculis.	E.	18 40 9.72	+ 0.40	+ .21	− .14	+ 1 43 0.39	
July 30.35								+ 1 43 0.74
	α Scorpii.	W.	18 4 39.95	+ 13.57	+ .05	+ 2.09	+ 1 42 56.16	
	x Ophiuchi.	E.	18 34 49.14	+ 5.72	+ .20	− 1.94	+ 1 42 56.32	
	α¹ Herculis.	E.	18 52 2.04	+ 4.57	+ .13	− 1.98	+ 1 42 56.45	
July 31.36								+ 1 42 56.31
	d Sagittarii.	E.	20 53 19.08	+ 13.70	− .24	− 2.46	+ 1 42 55.99	
	δ Cygni.	W.	21 24 11.96	+ 4.73	− 1.38	+ 3.21	+ 1 42 55.65	
	α Aquilæ.	W.	21 27 41.58	+ 6.52	− 1.12	+ 2.32	+ 1 42 56.73	
July 31.47								+ 1 42 56.12

There were also made the following —

CHRONOMETER COMPARISONS AT FORT WORTH.

Date. 1878.	Victor Kullberg, 1178, s. t.	Bond & Sons, 1058, m. t.	Johnson, 1436, m. t.	Observer.	Date. 1878.	Victor Kullberg, 1178, s. t.	Bond & Sons, 1058, m. t.	Johnson, 1436, m. t.	Observer.
July	h. m. s.	h. m. s.			July	h. m. s.	h. m. s.		
18.25	15 26 56.5	5 56 50.0		R. W. W.	24.40	19 31 39.5	9 38 10.0		R. W. W.
19.35	17 56 20.5	8 22 3.0		L. W.	24.94	8 32 44.5	22 37 12.0		L. W.
21.38	18 47 30.0	9 4 28.0		R. W. W.	25.08	11 55 58.0	1 59 53.5		R. W. W.
21.49		11 46 2.5	11 45 40.0	L. W.	25.38		9 33 0.0	9 31 34.0	L. W.
21.94	8 17 6.0	22 33 54.0		L. W.	25.40	19 29 26.5	9 32 10.5		L. W.
22.38	18 43 15.0	8 57 26.5		R. W. W.	26.10	12 26 21.0	2 26 24.0		R. W. W.
22.43		10 18 20.0	10 17 40.0	L. W.	26.39	19 18 47.0	9 17 45.0		R. W. W.
22.46	20 47 8.0	11 1 0.0		R. W. W.	26.41		10 3 10.0	10 1 30.0	L. W.
22.89	7 24 29.5	21 36 41.0		L. W.	27.42		10 18 30.0	10 16 37.0	L. W.
23.33	17 47 35.5	7 52 10.0		R. W. W.	27.98		23 25 0.0	23 23 0.5	L. W.
23.37		9 8 30.0	9 7 35.8	L. W.	28.93	8 22 49.0	22 12 10.0		J. K. R.
23.42	19 44 47.0	9 55 2.0		R. W. W.	28.99	9 58 0.0	23 47 5.0		J. K. R.
23.96	8 52 29.5	23 0 40.0		L. W.	29.02	10 39 25.0	0 28 24.0		L. W.
24.08	11 45 25.0	1 53 8.5		L. W.	29.02		0 36 30.0	0 35 13.0	L. W.
24.37		9 11 30.0	9 10 20.0	L. W.	29.41	20 3 36.0	9 51 6.5		R. W. W.
					30.42	20 11 0.0	9 54 42.5		J. K. R.

Whence we have the —

ADOPTED CHRONOMETER ERRORS AND RATES AT FORT WORTH.

Date.	Error "Kullberg, 1178."	Rate K. 1178.	Date, 1878.	Error "Bond, 1058."	Rate "Bond, 1058."	Date, 1878.	Error "Johnson, 1436."	Rate J. 1436.
	h. m. s.	s.	d.	s.	s.	d.	s.	s.
1878 July 20.37	+ 1 43 34.45	− 3.73	16.00	+ 6. 7*	+ 3.03	21.49	+ 8.2	12.6
22.43	+ 1 43 26.78	− 3.49	19.00	+ 16. 8*	+ 5.62	22.43	− 4.0	9.4
23.37	+ 1 43 23.50	− 3.44	22.46	+ 36.17	+ 5.78	23.37	− 12.8	10.1
25.37	+ 1 43 16.62	− 3.18	23.42	+ 41.71	+ 5.72	24.37	− 22.9	10.0
27.40	+ 1 43 9.16	− 1.89	25.40	+ 53.03	+ 5.09	25.38	− 33.0	7.3
28.46	+ 1 43 7.16	− 3.40	26.39	+ 58.06	+ 6.08	27.98	− 51.8	Set by 1058.
30.35	+ 1 43 0.74	− 4.47	28.93	+ 73.51	+ 5.55	29.02	− 2.9	
31.36	+ 1 42 56.31	− 1.8	30.42	+ 81.75				
31.47	+ 1 42 56.12		* Sextant Observations.					

THE LONGITUDE DETERMINATION.

———

THROUGH the courtesy of the officers of the Western Union Telegraph Company and of the Texas and Pacific Railroad Telegraph Company, arrangements were made to transmit longitude signals between Fort Worth and St. Louis on the nights of July 15th and July 31st. With this end in view, there was sent to Mr. Willson at Fort Worth the "Brown University" transit instrument and the mean-time box chronometer, "Bond, 1058." At St. Louis, I used the "Temple transit," and the sidereal box chronometer, "Victor Kullberg, 1178," on the evening of July 15th; and the mean-time box chronometer, "Dent, 2748," on the evening of July 31st. For the reason previously mentioned in the Introduction, however, an exchange on the night of the 15th proved impracticable; and we are therefore compelled to rely on the one night's exchange of signals of July 31st. The following programme for the observer at Fort Worth will give the method by which the exchange was effected : —

LONGITUDE PROGRAMME.

OBSERVING STATION AT FORT WORTH.

1878. July 15. Begin with Circle West.
    31.   „   „   „   East.

OBSERVING SCHEME.

1. Level and reverse level.
2. Observe five or six stars, two of which are circumpolar (*i. e.* $\delta > 65°$), observing for level near the middle of the series.
3. Level and reverse level.
4. Be at Telegraph Office at 9.40 Fort Worth mean time, and adjust relay so the *back* stroke will be strongest, and use this back stroke throughout the longitude work. Exchange the signals as in the scheme following.
    The 1st night the signals will begin from St. Louis, the 2d night they begin from Fort Worth.
5. Return to observing station, and reverse the telescope upon a circumpolar star for collimation, reading the level before and after reversal.
6. Observe five or six stars as in 2.
7. Level and reverse level as in 1.

TELEGRAPH OFFICE AT FORT WORTH.

1878. July 15. 9^h 45^m Fort Worth mean time.

Person using the Line.	Message or Signal.
L. W.	"Are you ready?"
R. W. W.	"Yes. Begin."
L. W.	Rattle at the next 45' beat of L. W.'s chronometer. Rattle continues from 45' to 55', and the chronometer beats

       0' 2   4   6   8 10 12 14 16 18 20 22 24
       30 32 34 36 38 40 42 44 46 48 50
        0   2   4   6   8' &c., are sent by tapping with the telegraph key for 5^m, the last signal being at

o' of the 6th minute. Rattle from 5' to 15' of the 6th minute completes the signal, and L. W. then telegraphs the hour and minute of the o' following the first rattle.

R. W. W.	"Are you ready?"
L. W.	"Yes. Begin."
R. W. W.	Signals precisely as above.
L. W.	"Are you ready?"
R. W. W.	"Yes. Begin again."
L. W.	Signals precisely as above except they continue for three minutes instead of five, and the last rattle occurs 4ᵐ 5' to 4ᵐ 15', hour and minute of first signal telegraphed as before.
R. W. W.	"Are you ready?"
L. W.	"Yes. Begin again."
R. W. W.	Signals as in the preceding clause.
R. W. W.	"Good-night."
L. W.	"Good-night."

Two break-circuit keys, precisely similar in design and size, were constructed; and one was used by Mr. Willson at Fort Worth, the other by myself at St. Louis.

Both observers having repaired to the Telegraph Office upon the night of July 31st, the following comparisons were made, in accordance with the above programme: —

I.

Time of sending signals from Fort Worth by "Victor Kullberg, 1178," s. t. — Time of receiving signal at St. Louis by "Dent, 2748," m. t.

h.	m.	s.		h.	m.	s.
19	56	30.0	=	10	6	27.9
	57	0.0	=		6	57.8
	57	30.0	=		7	27.8
	58	0.0	=		7	57.8
	58	30.0	=		8	27.5
	59	0.0	=		8	57.4
	59	30.0	=		9	27.3
20	0	0.0	=		8	57.3
	0	30.0	=		10	27.2
	1	0.0	=		10	57.1
20	1	30.0	=	10	11	27.0

II.

Time of receiving signals at Fort Worth by "Victor Kullberg, 1178." — Time of sending signals from St. Louis by "Dent, 2748."

h.	m.	s.		h.	m.	s.
20	5	3.5	=	10	15	0.0
	5	33.6	=		15	30.0
	6	3.65	=		16	0.0
	6	33.7	=		16	30.0
	7	3.85	=		17	0.0
	7	33.9	=		17	30.0
	8	4.1	=		18	0.0
	8	34.15	=		18	30.0
	9	4.2	=		19	0.0
	9	34.35	=		19	30.0
20	10	4.45	=		20	0.0

III.

Time of sending signals from Fort Worth by "Victor Kullberg, 1178."				Time of receiving signals at St. Louis by "Dent, 2748."		
h.	m.	s.		h.	m.	s.
20	13	0.0	=	10	22	55.1
		30.0	=		23	20.0
	14	0.0	=		23	55.1
		30.0	=		24	25.0
	15	0.0	=		24	54.9
		30.0	=		25	24.8
20	16	0.0	=	10	25	54.6

IV.

Time of receiving signals at Fort Worth by "Victor Kullberg, 1178."				Time of sending signals from St. Louis by "Dent, 2748."		
h.	m.	s.		h.	m.	s.
20	19	5.9	=	10	29	0.0
	19	36.1	=		29	30.0
	20	6.1	=		30	0.0
	20	36.15	=		30	30.0
	21	6.2	=		31	0.0
	21	36.3	=		31	30.0
20	22	6.45	=	10	32	0.0

The time observations made at Fort Worth by Mr. Willson and Professor Rees will be found as a continuation of the regular series made at Fort Worth. The transit used at St. Louis is a very nicely finished one, made by Temple, of Boston. It has an aperture of 1.7 inches, and a focal length of about 28 inches. The value of one division of the level I found to be

$$1^d.00 - 8''.68 \text{ at } 74°.8,$$

which is the mean of ten settings with the Meridian Circle of the Observatory of Harvard College.

The corrections necessary to reduce each of the five webs in the reticule to the mean web are as follows for Circle East: —

I.	II.	III.	IV.	V.
+ 46'.082	+ 23'.073	- 0'.051	- 22'.890	- 46'.482.

The St. Louis observations were very imperfect previous to going to the Telegraph Office since I was hardly able to get the instrument approximately into the meridian, and its level and collimation adjusted, before it was time to repair to the Telegraph Office.

The observations are as follows : —

Date. 1878.	Star's Name.	Lamp.	Observed Time. Dent, 2748. St. Louis m. t.	Aa.	Bb.	Cc.	Error, Dent, 2748.
			h. m. s.	s.	s.	s.	m. s.
July 31.4	μ Herculis *	E.	9 6 18.18	- 1.12	- 0.22	- 0.36	+ 2 54.88
	ψ¹ Draconis	E.	9 8 33.78	+ 9.68	- 0.57	- 1.05	+ 2 55.70
	γ Draconis *	E.	9 18 18.64	+ 1.93	- 0.39	- 0.51	+ 2 54.99
	τ Aquilæ *	W.	11 22 27.36	- 2.84	- 0.42	+ 0.32	+ 2 54.82
	κ Cephei	W.	11 36 54.63	+ 15.25	- 1.70	+ 1.47	+ 2 55.69
	ζ Cygni *	E.	12 31 43.36	+ 4.28	- 0.35	- 0.38	+ 2 55.12
	ι Pegasi *	E.	12 40 20.18	+ 8.32	- 0.35	- 0.35	+ 2 55.21
	β Cephei	E.	12 51 40.63	- 36.38	- 1.02	- 0.97	+ 2 53.63
	ε Pegasi *	E.	13 1 58.17	+ 11.89	- 0.44	- 0.33	+ 2 55.22

The observations show a disturbance of the instrument after the observation of κ Cephei, which at first was so irregular in its nature that several stars observed with lamp W have been rejected. Solving for a new azimuth after κ Cephei, and taking the mean of the time stars, which are denoted by a (*), we have, after reducing the errors to the same instant,* —

The error of Dent, 2748, July 31ᵈ.47, is + 2ᵐ 55ˢ.01 ± 0ˢ.03.

Now, since the signals were sent by hand, we shall arrive at a more probable comparison of the chronometers, if we take the means as the comparisons stand than if we use the coincidences only. Whence we have, —

	Victor Kullberg, 1178.				Dent, 2748.		
	h.	m.	s.		h.	m.	s.
From I.	19	59	0.00	=	10	8	57.46
II.	20	7	33.95	=	10	17	30.00
III.	20	14	30.00	=	10	24	24.94
IV.	20	20	36.17	=	10	30	30.00

and the final mean, which will be free from errors of time of transmission and armature time, —

St. L. m. t.	Victor Kullberg, 1178 (s. t.)				Dent, 2748 (m. t.)		
	h.	m.	s.		h.	m.	s.
1878. July 31ᵈ.44.	20	10	25.030	=	10	20	20.600

And we have for the errors of the two chronometers, —

For July 31ᵈ.44 St. Louis m. t.   =   July 31ᵈ.42 Fort Worth m. t.

		h.	m.	s.	
V. Kullberg 1178	=	+	1	42	56.22 (s. t.)
Dent 2748*	=	+	0	2	54.96 (m. t.)

Whence we have the final comparison, reduced to their respective meridians, but still affected by the personal equations of the observers, R. W. Willson and myself, —

	Fort Worth sid. time.				St. Louis mean time.		
	h.	m.	s.		h.	m.	s.
1878. July 31ᵈ.4.	18	27	28.81	=	10	17	25.64

Assuming the right ascension of the mean sun for St. Louis for this date to be 8ʰ 36ᵐ 53ˢ.33, and applying the difference of the personal equations of Mr. Willson and myself as determined by transit observations after our return to Cambridge, and found to be 0ˢ.27 additive to the observation of an equatorial star by Mr. Willson, we have then the difference of longitude between the observatory of Washington University, St. Louis, and the transit pier in the rear of Mr. S. W. Lomax's residence, near Fort Worth, Texas, is, —

0ʰ 28ᵐ 31ˢ.31

and if we apply the small correction necessary to reduce the position of the transit pier of the Washington University Observatory to the position of the transit pier occupied by Professor William Eimbeck in the St. Louis — Washington longitude campaign, as described by Professor William Harkness (Washington Astr. Obs. 1870, App. I.), and adopt the longitude of Professor Eimbeck's pier, as described in the paper mentioned, we have, —

Fort Worth Transit, west of Washington University Observatory	0ʰ 28ᵐ	31ˢ.31
Washington University Observatory, west of Professor Eimbeck's pier	0ᵐ	0ˢ.13
Professor Eimbeck's pier west of United States Naval Observatory at Washington, D. C.	0ʰ 52ᵐ	36ˢ.90

And finally, the longitude of our observing station at Fort Worth is west 1ʰ 21ᵐ 8ˢ.34 of the dome of the United States Naval Observatory at Washington, D. C.

---

* Hourly rate deduced from sextant observations July 27th and August 5th is + 0ˢ.061.

We desired to receive the Washington noon-signal on the morning of the 29th of July, in order to compare our chronometers with the Washington clock. Professor Rees, who had this matter in charge, embodies the result of our efforts in this direction in the following words: —

" In the signal received, a rattle announced the coming of the taps; then, after a pause of a few seconds, a tap was sounded for every second, till exactly at twelve o'clock Washington mean time a double-beat was tapped. The double-beat was continued to exactly one minute after twelve o'clock Washington time, then was changed into a single tap for each second, which lasted for several seconds, when a rattle was sounded, and the signals ended. We had then the means of two comparisons, one at twelve o'clock Washington mean noon, and another one minute later.

" July 29th, 1878, A.M. — Comparison of chronometers to show whether Bond, 1058, suffered any change of rate in the carrying to and from the telegraph office: —

Mean time, Bond, 1058.				Sidereal time, Kullberg, 1178.		
h.	*m.*	*s.*		*h.*	*m.*	*s.*
22	12	10.0	=	8	22	49.0

" After receiving the signals, we have —

Bond, 1058.				Kullberg, 1178.		
h.	*m.*	*s.*		*h.*	*m.*	*s.*
23	50	0.0	=	10	0	55.0

" The mean solar interval between the two comparisons is 1^h 37^m $50^s.0$, which gives, reduced to sidereal time, 1^h 38^m $6^s.07$. The noted sidereal interval is seen to be 1^h 38^m $6^s.0$. But 1^h 37^m $50^s.0$ is not the *true* solar interval of the comparisons, but is too great by the amount that 1058 gains in that interval. Since the rate of 1058 is $+ 5^s.55$, it would gain in the interval $0.4'$; hence, the true mean solar interval is 1^h 37^m $49^s.6 = 1^h$ 38^m $6^s.07$ sidereal time. The noted sidereal interval, corrected for the loss of Kullberg in the interval, gives 1^h 38^m $6^s.23$. Thus, we seem to have evidence of a small change of rate in 1058. If we consider, as we think we have a right to do, that the change occurred partly in coming down and partly in going back to our station, then the change that had occurred up to the time of the signals will be small.

" July 28.94. — Signals from Washington.

Washington.				Bond, 1058.		
h.	*m.*	*s.*		*h.*	*m.*	*s.*
0	0	0.0	=	10	40	8. (comparison not good.)
	1	0.0	=		41	9. (comparison better.)

" The signals were constantly broken in on by the agents along the railroad line, and I was very much disturbed.

" Assuming the error of Bond, 1058, at the time of comparison, to be $+ 1^m$ $13^s.57$, we compute the following: —

	West of Washington.		
	h.	*m.*	*s.*
Longitude from first comparison	1	21	5.57
Longitude from second comparison . . .	1	21	4.57

" On informing Mr. Todd, at Dallas, that I had not received a satisfactory set of signals, he answered that he had received uninterrupted ones, and would send me signals, using his sidereal chronometer.

" He preceded his signal by a rattle of a few seconds, and then sent a signal every ten seconds by his chronometer, the time of reception being noted on Bond, 1058. The signals closed by Mr. Todd sending me the time of beginning, and his error on Washington sidereal time.

" The following is the record : —

Bond, 1058.				Washington Sidereal Chronometer.		
h.	m.	s.		h.	m.	s.
23	0	21.5	=	8	59	0.0
		31.0	=			10.0
		41.5	=			20.0
		51.0	=			30.0
		61.0	=			40.0
	1	10.5	=			50.0
		20.6	=	9	0	00.0

" Reducing all these comparisons to the same instant, we get : —

h.	m.	s.		h.	m.	s.
23	0	21.5	=	8	59	0.0
		21.03	=			
		21.56	=	[Mr. Todd's chronometer was		
		21.08	=	9^m 49^s.3 fast.]		
		21.11	=			
		20.64	=			
		20.76	=			

" Taking the mean, applying the Bond and sidereal chronometer corrections, and reducing Washington sidereal time to the corresponding mean time, we have : —

Washington mean time.				Fort Worth mean time.		
h.	m.	s.		h.	m.	s.
24	20	15.79	=	22	59	7.41

Hence the difference in longitude is, by the Washington signal : —

h.	m.	s.
1	21	8.38."

THE OBSERVATIONS FOR LATITUDE.

THE following observations were made with the sextant "Stackpole and Brother, 1707," by myself. The sextant is of the pattern originally designed by Professor William Harkness, and described by him in the "United States Naval Observatory Eclipse Report," of 1869; with the exception that it has an arc of seven inches radius instead of six inches, as there mentioned. It was received directly from Mr. Stackpole's hands, and was found to be in excellent adjustment throughout.

A method of determining the eccentricity of a sextant, where the conveniences of a large circle are at hand, occurred to me, and I have adopted it in the following investigation ; using the large Meridian Circle of the Harvard College Observatory for this purpose, through the courtesy of Professor W. A. Rogers.

I replaced the finding level on the index-arm of the sextant by another, so sensitive that an inclination of 10″ would displace the bubble about 0.03 inches. The sextant was then firmly secured to the meridian circle, so that the plane of the sextant should be, as nearly as practicable, parallel to the plane of the circle.

It then became a simple matter to set the vernier of the sextant at consecutive 10° graduations of the sextant-limb; and the mean readings of the meridian circle, when the sextant-level is brought to the same position by moving the circle in declination, gave the true intervals between the zero of the sextant-limb and the points at which the vernier was successively set.

Putting $\Delta s =$ the sextant correction for any reading $2z$,

$E =$ the number of degrees from the diameter of maximum eccentricity to the zero of graduation of the sextant, counted in the direction of the graduation,

$e =$ the maximum eccentricity expressed in seconds of arc,

we have by the usual formula, with all necessary accuracy, an equation of the form —

$$\Delta s = e \sin (z + E)$$

for each interval determined.   Also,

$$\Delta s = e \cos E \sin z + e \sin E \cos z,$$

whence we form the normal equations, —

$$\Sigma \sin z \, \Delta s = \Sigma e \cos E \sin {}^2z + \Sigma e \sin E \sin z \cos z ;$$
$$\Sigma \cos z \, \Delta s = \Sigma e \cos E \sin z \cos z + \Sigma e \sin E \cos^2 z.$$

Proceeding in the manner indicated, I found for the sextant under consideration, —

$$e = 22''.1, \qquad\qquad E = 14°.0 ;$$

whence we derive the following table of corrections to any observed angle $z :$ —

z	Δz	z	Δz
°	''	°	''
0	− 5.35	40	− 17.87
10	− 8.99	50	− 19.88
20	− 12.35	60	− 21.23
30	− 15.35	70	− 21.99

Referring these angles to the sextant-limb, and making the correction 0'' at the zero of the graduation, we have, denoting the observed sextant reading for an angle measured by ω, —

ω	$\Delta\omega$	ω	$\Delta\omega$
°	''	°	''
0	− 0.0		
10	− 1.9	70	− 11.3
20	− 3.6	80	− 12.5
30	− 5.4	90	− 13.6
40	− 7.0	100	− 14.5
50	− 8.7	110	− 15.3
60	− 10.0	120	− 15.9

In the observations for latitude, the fourth column indicates the reversal of the artificial horizon cover, I. and II. being the ends respectively nearest the observer. The fifth column indicates the number of double altitudes corresponding to the mean of the times and double altitudes given in columns six and seven. Column eight contains the correction to my pocket chronometer, "Johnson, 1436," in Fort Worth local mean time. Column nine contains the index correction, which always depends upon from six to ten observations made either between the sets of double altitudes taken in reversed positions of the horizon glass cover, or else immediately following the double altitude observations.

An eye-piece magnifying about seven diameters was used, and after the date of July $24^{s}3$ a stand was employed to support the sextant. Mr. Seagrave recorded the observations at my dictation.

The resulting latitudes in the case of α Ursæ Minoris were obtained by using the tables published in the supplement to the "American Ephemeris and Nautical Almanac for 1878," for the factors A, B, C, D. Since the hour angle of the star is nearly 6^{h} in the observations made below, we shall not introduce sensible error by assuming the motion of α Ursæ Minoris in altitude to be proportional to the time for the short intervals used. Accordingly, the observations have been taken in sets of five, and their means used in the reductions. In the case of the circum-meridian altitudes of α Scorpii, since the zenith distance is $67°$, we shall not require to consider the remaining terms of the formula (following Chauvenet's notation), —

$$h_1 = h_0 + Am_0 \qquad \text{in which}$$

h_0 the mean of the observed altitudes corrected for index error, eccentricity, and refraction.

$A = \cos\phi \cos\delta \operatorname{cosec} \zeta_1.$

$\zeta_1 =$ the assumed zenith distance.

$m_0 =$ the mean of the values of m.

$m = \dfrac{2 \sin^2 \frac{1}{2} t}{\sin 1''},$

where t is expressed in sidereal seconds, and indicates the distance from the meridian.

And, accordingly, the resulting latitudes have been computed by the formula, —

$$k_1 - k_0 + Am_0 .$$

The assumed position of the observing station is

$$\lambda = 1^h \ 21^m.0 \quad \text{W. of Washington,}$$
$$\phi = 32^\circ \ 44'.4 \quad \text{N.;}$$

and the star places adopted are —

1878 July 25^d.4   a Ursæ Minoris   1^h 14^m 16^s.6   + 88° 39′ 29″.2
              a Scorpii   16 21 59 .6   − 26   9 49 .4

SEXTANT OBSERVATIONS OF a SCORPII FOR LATITUDE.

Date.	Bar.	Ther.	Horizon	No. Obs.	Chronometer "Johnson, 1436."	Sextant "Stackpole, 1707"	Δc, "J. 1436."	Index.	δ	Resulting Latitude.
	in.	*°*			*h. m. s.*	*° ′ ″*	*m. s.*	*′ ″*	*″*	*° ′ ″*
1878 July 23.34	29.28	81.7	I	1	8 0 25	62 4 15	+ 0 12.4	− 1 56	− 10	
			I	1	8 2 10	62 6 35				
			I	1	8 4 20	62 9 10				
			I	1	8 5 17	62 10 0				
			I	1	8 6 33	62 12 0				
			I	1	8 7 50	62 14 0				
			I	1	8 8 52	62 13 50				
			I	1	8 11 17	62 14 20				
			I	1	8 12 14	62 15 20				32 44 21.6
			I	1	8 15 16	62 15 50				
			II	1	8 17 49	62 15 50				
			II	1	8 18 48	62 15 10				
			II	1	8 20 43	62 14 0				
			II	1	8 22 10	62 12 20				
			II	1	8 24 3	62 11 30				
24.34	29.12	83.0	I	1	8 10 50	62 16 30	+ 0 22.6	− 2 0	− 10	
			I	1	8 12 12	62 16 0				
			I	1	8 14 8	62 15 0				
			I	1	8 15 7	62 14 20				
			I	1	8 16 5	62 15 0				
			I	1	8 16 58	62 14 5				32 44 25.9
			II	1	8 19 33	62 12 0				
			II	1	8 21 20	62 10 0				
			II	1	8 25 30	62 3 0				
			II	1	8 26 45	62 1 10				
25.33	29.00	82.0	I	1	8 1 32	62 15 0	+ 0 32.5	− 3 3	− 10	
			I	1	8 5 50	62 16 10				
			I	1	8 6 55	62 16 26				
			I	1	8 9 26	62 16 0				
			I	1	8 12 38	62 15 30				32 44 45.9
			II	1	8 16 20	62 11 35				
			II	1	8 17 57	62 10 0				
			II	1	8 20 30	62 6 10				
			II	1	8 22 10	62 4 0				
			II	1	8 23 13	62 1 45				

SEXTANT OBSERVATIONS OF α URSÆ MINORIS FOR LATITUDE.

Date.	Bar.	Ther.	Horizon	No. Obs.	Chronometer "Johnson, 1436."	Sextant "Stackpole, 1707"	Al. J. 1436.	Index.	e.	Resulting Latitude.
d.	in.	"			h. m. s.	° ' "	m. s.	' "	"	° ' "
1878 July 21.46			I	5	11 1 57.1	65 24 50	- 0 8.5	- 2 37	- 11	32 44 35.2
21.47			II	5	11 16 26.6	65 35 3	- 0 8.4	- 2 37	- 11	44 13.9
22.41	29.28	80.7	I	5	9 51 47.8	64 38 30	+ 0 4.5	- 2 31	- 10	44 8.3
22.42			II	5	9 59 59.6	64 45 4	+ 0 4.5	- 2 31	- 10	44 42.8
23.37	29.28	81.0	I	5	8 43 31.6	64 0 5	+ 0 12.8	- 2 29	- 10	44 41.7
23.38			II	5	8 52 4.6	64 4 19	+ 0 12.9	- 2 29	- 10	44 21.7
24.37	29.34	83.0	I	5	8 44 8.0	64 1 45	+ 0 22.9	- 2 29	- 10	44 30.1
24.38			II	5	9 5 1.2	64 13 20	+ 0 23.0	- 2 29	- 10	43 51.9
25.37	29.00	82.0	I	5	8 47 11.6	64 5 56	+ 0 33.0	- 3 43	- 10	43 57.9
25.38			II	5	8 59 8.0	64 13 30	+ 0 33.1	- 3 3	- 10	44 10.1
26.38	29.04	81.0	I	5	8 56 22.6	64 14 15	+ 0 40.3	- 2 47	- 10	44 32.0
26.39			II	5	9 16 10.6	64 26 3	+ 0 40.4	- 2 47	- 10	32 44 11.8

The observations of α Ursæ Minoris are more numerous than those of α Scorpii; but since the observations of α Scorpii were made after the stand was in use, and I had become better accustomed to the insects which, attracted by the light, caused considerable annoyance during the observations, I shall give equal weight to each of the two stars; and we have, from star observations, the latitude of the transit pier of our observing station is, —

By 60 Double Altitudes of α Ursæ Minoris 32° 44′ 18″.8
By 35 Double Altitudes of α Scorpii   32 44 31 .1

Whence we have —

$$\phi = 32° \; 44′ \; 25″.$$

The following sun observations were made and reduced by Professor Rees at my suggestion : —

Civil Date. 1878.	Corrected Bar.	Therm.	Number Obs.	Means of Bord, 1058, A.M.	Means of ☉ 2 alt.	Means of Bord, 1058, P.M.	Index Correction.	Eccentricity.	Deduced φ.
		F.		h. m. s.	° '	h. m. s.	' "	"	° ' "
Sat. July 20	29.56*	96*	6	10 34 5.3	132 20		- 2 23.1	- 16.3	} 32 44 10.90†
,, ,, 20	29.56*	96*	5	††	129 0	1 46 8.0	- 2 14.9	- 16.3	}
Mon. ,, 22	29.36	96	20	10 22 56.71	127 30	1 50 28.76	- 2 23.0	- 16.3	32 45 19.
Tues. ,, 23	29.34	91	10	10 24 5.79	127 40		- 2 33.0	- 16.3	} 32 44 53.8†
,, ,, 23	29.24	92.2	10		126 30	1 52 38.59	- 2 41.3	- 16.3	}
Wed. ,, 24	29.25	100.8	20	10 25 15.2	127 50	1 48 38.74	- 2 34.4	- 16.3	32 45 44.
Thurs. ,, 25	29.11	98.8	24	10 26 40.47	128 5	1 47 28.08	- 2 34.15	- 16.3	32 45 52.6
Total Observations			95					Mean	32 45 12.06

* Assumed.
† These have been deduced by using the formulæ for single altitudes of the sun and taking their mean (Chauvenet, Vol. I. p. 230, University edition).
†† This observation is of the sun's lower limb.

Although the observations agree well *inter se*, yet I am apprehensive there is a systematic error arising from some or all of the following causes : —

1° The intense heat would affect the horizon, the sextant, and the observer. We have no means of determining the effect of heating the sextant so hot as it became in the course of these observations, and I am of the opinion that the results would not be uniform for consecutive days.

2° The sun's hour angle was so great that small errors of observation would be far more perceptible than when the sun was near the meridian.

3° The same thing which obliged Professor Rees to observe the sun at a larger hour angle than desirable also obliged him to use the extreme limits of the sextant graduation.

4° The chronometer rate might be irregular under the influence of the heat, and the hour angles be in error for that reason.

It was desirable, however, that solar observations should be made ; and, until the stand came into use for the star observations, it was highly probable that the solar observations would be better than the stellar observations. I am now of the opinion, however, that we shall have a more accurate determination of ϕ if we use the value deduced only from the observations of a Scorpii and a Ursæ Minoris.

GEOGRAPHICAL POSITIONS.

POSITION OF FORT WORTH, AND THE APPROXIMATE POSITIONS OF THE TOWNS MENTIONED IN THE REPORTS OF OTHER OBSERVERS.

In order to connect our observing station with some permanent monument in the City of Fort Worth, the City Engineer, Mr. J. C. Terry, made the following survey, in which "station 1 " designates the wooden transit pier in the rear of Mr. Lomax's residence, and "station 8 " designates the stone monument in the Court House Square : —

No. of Station.	Angle.			Course.				Distance. Feet.
1	0°	0′		North.				200
2	46°	30′	Right.	N.	46°	30′	E.	1700
3	22°	0′	„	N.	68°	30′	E.	500
4	21°	30′	„	East.				210
5	90°	0′	Left.	North.				1330
6	29°	45′	„	N.	29°	45′	W.	2800
7	90°	0′	Right.	N.	60°	45′	E.	400
8	56°	31′	„	S.	63°	15′	E.	130

From these data, Mr. Terry finds the monument to be, —

North 5454.3 feet, and east 982.1 feet of the transit pier.

If now we assume that in latitude 32°45′ north, 60″ measured on the parallel is 5152.2 feet and 60″ measured on the meridian is 6063.6 feet, we have the position of the monument with reference to the transit pier to be, —

In latitude + 53″.7   In longitude − 11″.5 = − 0ˢ.77,

whence the position of the monument in Court House Square is, —

Latitude north   32° 45′ 19″

Longitude west of the centre of the dome of the United States Naval Observatory at Washington, D.C.,

h. m. s.
1 21 7.57.

Mr. D. P. Todd, having made an examination of the existing maps of Texas, kindly informed me that the map issued by the Houston and Texas Central Railroad Company is as nearly correct as any he had seen. I have therefore, as the best available means of locating the observers co-operating with us, measured upon this map the following positions : —

Place.	Longitude, West of Washington.					Latitude North.	
	°	′	h.	m.	s.	°	′
Hearne	19	32.0 =	1	18	0	30	56.6
McKinney	19	33.1 =	1	18	12	33	14.1
Allen	19	36.3 =	1	18	25	33	6.1
Bremond	19	38.6 =	1	18	34	31	10.6
Dallas	19	43.4 =	1	18	54	32	48.6
Fort Worth	20	15.5 =	1	21	2	32	46.0

If we assume there is a systematic error in these positions equal to the error in the displacement of Fort Worth, we have a systematic correction of —

$$+ 5.6' \text{ in Longitude,}$$
$$- 41'' \text{ in Latitude,}$$

to be applied to each of the positions given above.

REPORTS.

―――――•―――――

REPORT OF MR. LEONARD WALDO.

―――――

IN the division of labor among ourselves for the observations on the day of the eclipse, it seemed expedient that I should become responsible for the photographic records made during totality. This, with that general oversight over all the observations which it was necessary for some one to assume, in order that our work should be complete and symmetrical as a whole, was thought to be quite as much as could be undertaken by one observer. I shall preface the discussion of our photographs, however, with the mention of a few miscellaneous observations I found it convenient to make.

TELESCOPIC AND OTHER OBSERVATIONS.

A few moments previous to the first contact, I placed a blanket on the ground, and arranged over it a camera tripod in such a manner that, lying flat on my back, I could without exertion have an easy gaze at the sun. I then requested Professor Alexander Hogg, of the Texas Agricultural and Mechanical College, to be ready at a given signal, which we had previously rehearsed, to stop the stop-watch I had set with my pocket chronometer, "Johnson, 1436," and compared as follows : ―

Johnson, 1436.				Stop-watch.	
h.	m.	s.		m.	s.
3	7	0.0	—	0	0.0

At $3^h 8^m$, I lay down upon the blanket, and placed an umbrella so that I was for the most part protected from the sun's intense rays, but still had full opportunity to point my hand-telescope of 1.35 inches' aperture upon the sun. I had adjusted to this telescope a terrestrial eye-piece magnifying 24.5 diameters, and a neutral tint shade-glass sufficiently dark to occasion no uneasiness of the eye after several minutes' gazing at the sun.

I now rested the object-glass end against the camera tripod, and examined the sun's limb at the expected position angle of first contact. The definition was superb. The edge was hard, sharp, and the curve unbroken. The mottled appearance was well marked. There was not a spot visible. As the computed time of contact drew near, I directed my attention more closely to the computed position angle of contact on the sun's limb, and presently recognized a sharp deviation from the convex curve which had previously bound the sun's limb.

This phase was recorded by the stop-watch as 4^m $25\frac{5}{6}'$, which would be by Johnson, 1436, 3^h 11^m $25'.5$.

My note at the time was that this "contact may be $2'$ late." But after seeing how the fourth contact appeared in my telescope, and comparing the last instant I was able to detect a variation in the sun's outline at the fourth contact with the instant at which this outline was as much broken as it was when I observed the first contact, I am confident that $12'$ would be none too large an estimate of lateness. I would give, therefore, the time of first contact to be 3^h 11^m $13'.5$; and I think this is liable to an uncertainty of $2'$. Immediately after this observation, I made the following comparison : —

Johnson, 1436.				Kullberg, 1178.		
h.	*m.*	*s.*		*h.*	*m.*	*s.*
3	15	54.0	—	13	28	41.0

The error of Johnson, 1436, was — $5'.2$ on Fort Worth mean time, and we have the Fort Worth mean time of first contact $= 3^h$ 11^m $18'.7 \pm 2'$.

The second and third contacts I could not observe without conflicting with other duties. The fourth contact was observed in the same manner as the first, with the exception that I used the beat of my pocket chronometer Johnson, 1436, instead of the stop-watch, which was inadvertently stopped by Professor Hogg before the contact took place.

For some unexplained reason, Johnson, 1436, had stopped previous to the second contact. I reset it immediately, and compared it after the fourth contact as follows : —

Johnson, 1436.				Bond & Sons, 1058.		
h.	*m.*	*s.*		*h.*	*m.*	*s.*
5	25	0.0	—	5	25	59.4

and I observed the last contact at 5^h 19^m $30'.0$ by Johnson, 1436. Whence we have the Fort Worth mean time of fourth contact $= 5^h$ 19^m $14'.5$, making due allowance for the error of Bond, 1058.

Immediately previous to the commencement of totality, I stood with my back to the sun, looking at the face of a stop-watch, and waiting for the final disappearance of the crescent of light. In close proximity to this phase, but whether $2'$ before or not I cannot say, I glanced toward the ground, my attention drawn by a fluttering motion of something, I did not know what. Every eye but mine was toward the eclipsed sun. I quickly removed the cap of the camera-box to make an exposure, and again looked at the ground away from the eclipsed sun. What I had suspected before I was now certain of. There were flitting shadows moving swiftly toward the east and south-east. They seemed to me regular in their occurrence, both as to time and distance apart; but how far apart they were or how many occurred in a second I cannot form an estimate, other than that they seemed like dark crests to waves, and might be a hundred feet apart, occurring three in a second. Of the existence of this phenomenon I am positive; but the estimates of magnitude and frequency of these shadows should be received with the greatest caution. I suppose this to be an observation of the shadows ordinarily seen at the second contact of total solar eclipses.

During the progress of the eclipse, I examined the sun's disc with Professor Rees's telescope, with a diagonal reflector of plane glass and various eye-pieces. My conclusions were as follows:—

1° The faculæ are seen clearly up to within say 2″ of the moon's limb. I *think* they are sharp to the very edge. The moon's limb has something of the "boiling" appearance; and I am not *sure* that the sun's photosphere is sharply defined in the jagged edges of the moon's advancing limb.

2° There seems to be a minute point of light separated from, but in the line of, a prolongation of one of the sun's cusps. (This was after second contact.) I do not know whether this is a phenomenon similar to that which produces "Bailey's Beads" or not.

The Photographic Work.

1° *Preliminary Preparations and Description of the Cameras.*

Guided by own experience, and strengthened in my opinion by the kindly advice of Mr. D. C. Chapman, who is well known as the assistant who has aided Mr. Rutherfurd in the production of his exquisite astronomical photographs, I thought that we could find some local photographer who would do the necessary work under proper supervision. We were not able to secure the use of a telescope properly corrected for the chemical rays, and so we depended upon the use of large cameras instead. It occurred to me that we might use an improvised mounting of the following description for these cameras to supply the place of a clock-work motion. The bases of the cameras rested upon a board, one end of which was connected by a hinge to a square wooden plunger which moved vertically in a smooth fitting wooden casing which was firmly secured to a post. To the eastward of this post another post was placed, so that the board, hinged to the plunger at its west end, would rest its east end upon the second post, and be free to slide on this post as its west end followed the plunger in its vertical motion in the cylinder. On the top of their respective boards, the cameras were so arranged that motion in azimuth could be communicated to them by means of a screw controlled by a cord held in the hand. The mode of manipulating the apparatus was as follows: the vertical box being filled with dry sand, an adjustable opening in its base was sufficiently opened to allow the descending plunger to follow the upper surface of sand on which it rested, with the angular velocity (measured with a radius equal to the distance from the second and eastern post to the lower end of the plunger) of the sun in altitude. Then it was hoped there would be no difficulty in giving the necessary slow motion by hand to the large screw-head to counteract the sun's motion in azimuth. As the day of the eclipse drew near, however, it became apparent that there was so much to be done that our time of preparation was all too short. The storm of July 27th and 28th completely prevented any trial of our improvised mountings, and only after the commencement of the eclipse on the 29th was the sun in such a position that it was possible to experiment upon the rate of the two motions. As totality drew near, I found that it would be impracticable to make more than a rough adjustment of either motion in the time remaining; and so I decided to give up the azimuth motion altogether, and allow the plunger to follow the outflowing sand, with the idea that the motion of the sun in a straight line on the plate would present fewer difficulties in determining the position angles of the prominences, than if there was an undetermined displacement in both altitude and azimuth.

As for the camera lenses themselves, they were as follows: —

1° A Dallmeyer lens described as 6.D. The back combination only of this lens was used, which is of 48 inches focus.

2° A Ross Cabinet portrait lens, No. 3.

3° A Tench portrait lens of 6 inches' focus and 2.75 inches' aperture.

Between the lenses of 2° there was inserted a compound prism for polarizing light, made by Duboscq. This prism is composed of two bi-refracting prisms, so turned as to combine the refraction of each: they are achromatized with prisms of crown glass. The axial ray is the extraordinary one. The longer diagonal of the rhomb was nearly vertical, consequently the two images would be nearly horizontal.

Between the lenses of 3° there was inserted an ordinary double refracting prism with the shorter diagonal of the rhomb nearly vertical, consequently the two images would be nearly vertical. Both of these prisms were loaned by Dr. A. Lytton, of Washington University.

These last two cameras were thus provided at the suggestion made by Professor E. C. Pickering in his report to the superintendent of the United States Coast Survey upon the observations of the total solar eclipse of Dec. 22, 1870.

At four o'clock, Mr. Freeman, our photographer, retired to the dark room, and prepared three plates, one for each camera; and at my signal, about two minutes before totality, he placed them in their respective cameras, and again retired to the dark room. Two baths were used in coating the plates, and Mr. Freeman at once proceeded to prepare two more plates. During totality, the exposures were made without mishap, and the results are tabulated below: —

Number of the Negative.	Number of the Camera.	Time of Commencement of Exposure, counted from the Beginning of Totality.	Duration of Exposure.	Remarks.
1	1°	1′	60′	Shows 3 striæ. Coronal extension, $p = 120°$ is 23′; for $p = 300°$ is 25′.
2	1°	62′	65′	Shows the 3 striæ of No. 1 and 3 additional ones. Coronal extension, $p = 120°$ is 25′; for $p = 300°$ is 23′.
3	1°	130′	17′	Shown in Plate I., and is the best of the three negatives. Shows the 6 striæ of No. 2, and additional ones.
4	2°	2′	145′	Images badly defined and over-exposed.
5	3°	2′	145′	Images fairly sharp.

Owing to the short exposure of negative No. 3, the details are less obliterated than in the case of Nos. 1 and 2. Mr. S. B. Wells, of Boston, very kindly made for us a somewhat enlarged copy of this negative, which was afterwards still further enlarged by Mr. Edwards of the Heliotype Company, Boston, and is shown in Plate I. The negative was taken with a Tolles 4-inch objective and extremely oblique light: as a consequence, there are visible scratches on the glass plate, which with direct illumination would not have been visible. The unfortunate motion which is shown in Plate I., in negatives 1 and 2, is much more conspicuous, owing to their longer exposure. It will be noticed that the images of the sun's edge at the commencement and the end of the exposure are fairly outlined; and it will suffice for our necessarily approximate investigation of the details shown to consider these outlines as circular, and having a radical axis which is better determined from negatives 1 and 2 than from 3. This axis is inclined to the horizon in negative No. 1, 33°.1, and in negative No. 2, 32°.2. Adopting the mean, and assuming the exposure of No. 3 to have taken place under similar conditions to the exposure of Nos. 1 and 2, it would seem that the radical axis of No. 3 is inclined to a vertical circle 57°.3, counting toward the north, at the time and place of its exposure. The position angle measured by a vertical circle passing through the centre of the sun at the middle of totality at Fort Worth would be 61°.7; hence the position angle of the radical axis is 4°.4.

To locate as far as possible the striæ which are plainly evident in the area of the photograph comprised in the lunes of the intersecting circles, I measured upon the original negative the rectangular co-ordinates of the terminal points on the circumference of each circle. These measures were made under a magnifying power of fifteen diameters, obtained with a Tolles 4-inch objective and a micrometer eye-piece by Powell and Leland fitted to my microscope micrometer. (For a description see Proc. Am. Acad. Arts and Sci., 1877, p. 352.) Discussing the two images separately, each will afford us a determination of the angular intervals separating the terminal points of these striæ.

The position of each point may be expressed by an equation of the form —

$$x^2 + y^2 - r^2 - Ax - By + C = 0,$$

where x and y are the rectangular co-ordinates of the centre of the first image, and r its radius. A, B, are twice the measured co-ordinates of the terminal points, and —

$$C = \tfrac{1}{4}\,(A^2 + B^2).$$

In the same manner for the second image, we have —

$$x'^2 + y'^2 - r'^2 - A'x' - B'y' + C' = 0.$$

Tabulating the measured values of $\tfrac{1}{2}$ A, $\tfrac{1}{2}$ B, and $\tfrac{1}{2}$ A', $\tfrac{1}{2}$ B', we have the following table in which the terminal point given in the second image is not necessarily the same point mentioned in the first column, except in the case of the intersections of the two outlines: —

Designation of the point measured.	FIRST IMAGE.			SECOND IMAGE.		
	$\tfrac{1}{2}$ A.	$\tfrac{1}{2}$ B.	Remarks.	$\tfrac{1}{2}$ A'.	$\tfrac{1}{2}$ B'.	Remarks.
Intersection	10.3	0.7		10.2	0.0	
Terminal point	19.2	4.4	Well defined.	21.4	4.1	Well defined.
,, ,, ...	20.1	6.0	Faint.	22.5	5.4	Faint.
,, ,, ...	21.1	7.8	Faint.	23.3	7.5	Faint.
,, ,, ...	21.4	8.7		23.4	8.3	Beginning of the black
,, ,, ...	21.5	10.8	Beginning of the black	23.7	11.2	edge.
,, ,, ...	21.5	12.2	Hazy. [edge.	23.9	11.0	

Designation of the point measured	FIRST IMAGE.			SECOND IMAGE.		
	½ A.	½ B.	Remarks.	½ A'.	½ B'.	Remarks.
Terminal point	21.3	14.1	Faint.	23.9	11.9	
" "	20.8	15.8	Broad and hazy.	23.3	14.0	
" "	20.0	17.1		23.0	15.4	
" "	19.2	18.3	Well defined.	21.1	18.0	
Intersection	13.1	21.7		13.2	22.1	
Terminal point	5.3	20.9		7.3	20.2	Hazy at edges.
" "	3.2	19.1		5.2	18.9	Faint.
Point in the circumference .	2.6	18.2		4.0	17.0	
" " .	0.4	14.5		2.8	14.8	
" " .	0.0	9.8		2.0	11.0	
" " .	1.1	6.2		3.3	6.1	
" " .	3.9	2.7		5.6	3.0	

These data give us nineteen equations of condition for each image, involving x, y and their second powers. Forming the equations, adding them together, and taking their means, we have for the two cases —

$$x^2 + y^2 - r^2 - 25.90\,x - 24.08\,y + 425.91 = 0.$$
$$x'^2 + y'^2 - r'^2 - 29.80\,x' - 23.16\,y' + 470.56 = 0.$$

Subtracting each equation in turn from our mean equations, we shall then be rid of the second powers, and may proceed at once to form our normals with all necessary exactness, by rendering the coefficients of x, y, x', y', successively positive, and adding the resulting series of equations. We shall thus have —

$$307.10\,x + 18.56\,y - 3509.07 = 0 \quad \text{and} \quad 322.00\,x' - 35.96\,y' - 3773.74 = 0.$$
$$-8.30\,x + 206.72\,y - 2244.49 = 0 \qquad\qquad -41.60\,x' + 191.56\,y' - 1561.06 = 0.$$

Solving these equations for x, y, x' and y', and substituting in the mean equations to determine r and r', we have —

$$x = 10.796 \qquad x' = 12.943.$$
$$y = 11.289 \qquad y' = 10.954.$$
$$r = 10.883 \qquad r' = 10.893.$$

With these values of the co-ordinates of the centres of the two images, we may now easily refer each terminal point measured to the centre of its image. We have —

X	Y	θ	X'	Y'	θ'	X	Y	θ	X'	Y'	θ'
*− 0.5	−11.0	267.4	*− 2.6	−10.7	256.3	+10.5	+ 2.8	14.9	+11.0	+ 0.9	4.7
+ 8.4	− 6.9	320.6	+ 8.5	− 6.9	320.9	+10.0	+ 4.5	24.2	+10.4	+ 3.0	16.1
+ 9.3	− 5.3	330.3	+ 9.6	− 5.6	329.7	+ 9.2	+ 5.8	32.2	+10.1	+ 4.4	23.5
+ 10.3	− 3.5	341.2	+ 10.4	− 3.5	341.4	+ 8.4	+ 7.0	39.8	+ 8.2	+ 7.0	40.5
+ 10.6	− 2.6	346.2	+ 10.5	− 2.7	345.6	*+ 2.3	+10.6	77.8	*+ 0.2	+10.9	88.9
+ 10.7	− 0.5	357.3	+ 10.8	− 0.2	358.9	− 5.5	+ 9.6	119.8	− 5.6	+ 9.2	121.3
+ 10.7	+ 0.9	4.8	+ 11.0	− 0.0	360.0	− 7.6	+ 7.8	133.3	− 7.7	+ 7.9	134.3

* Radical axis.

The origin has been taken at right angles to the radical axis. The correction to the angle of the radical axis is —

$$\text{Sin } \Delta p = \pm \sqrt{\frac{(x - x')^2 + (y - y')^2}{4\,r^2}}$$

which gives —

$$\Delta p = \pm 5°.72.$$

By an examination of the photograph and its position with reference to the camera, it has been concluded that the point of the radical axis first mentioned is that point having a position angle of 184°.4. Applying the correction Δp and making this assumption, we have the following results: —

FIRST IMAGE.		SECOND IMAGE.	
Corrected Position angle.	Remarks.	Corrected Position angle.	Remarks.
184.4	Diameter parallel to radical axis.	184.4	Diameter parallel to radical axis.
243.3		243.3	
253.0		252.1	
263.9		263.8	
268.9		268.0	
280.0		281.3	
. . . .	Too faint to measure.	282.4	
287.5		287.1	
297.6		298.5	
306.9		305.9	
314.9		. . . .	Too faint to measure.
322.5		322.9	
6.2	Diameter parallel to radical axis.	5.6	Diameter parallel to radical axis.
42.5		43.7	
56.0		56.7	

The two columns give the position angles of the prominences as recorded on the negative No. 3 and reproduced on an enlarged scale in Plate I. These results are uncorrected for refraction, and their apparent discrepancy will be readily understood by those observers who have had occasion to measure photographic images of the sun. They are also liable to a small systematic correction, owing to the imperfect means of determining the zero of position.

I have no means of knowing whether these striæ represent prominences, strictly speaking, or whether they denote a greater intensity of coronal light only at the points whose position angles are given above. In Plate I. the details are fairly given, the extension of the corona particularly. The original negative has a lunar outline of about 0ᵐ.43 by direct measurement, and from this an idea of the amplification in the Plate may be obtained.

Negatives numbered 4 and 5 were submitted to Professor Pickering, who has kindly embodied the result of his examination of them in the following memorandum: —

"Each plate contains two photographs of the corona, formed by placing a double image prism between the lenses of the camera. One of these photographs is well defined (in negative No. 4), while the other is elongated in the direction of the line connecting them. Probably one image of the prisms is nearly achromatic, the dispersion of the other being uncorrected. The elongation is so marked in negative No. 4 that none of the differences can with certainty be ascribed to polarization.

"In negative No. 5, the images are nearer together, better defined, and the effect of color is much less.   The points marked *a*, *b*, are equally dark in the upper photograph, while the corresponding points *a'*, *b'*, of the lower photograph differ greatly in intensity, *b'* being much darker than *a'*.   This result is confirmed by *c'*, *d'*, which are much darker than *e'*, while *e* is but little lighter than *c* or *d*.

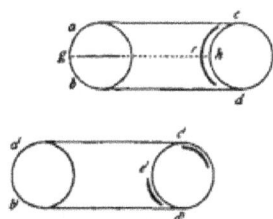

"These effects are explained by a radial polarization of the corona, if the plane of polarization of the achromatic image of the prism is perpendicular to the line connecting the two images.   Otherwise, they suggest tangential polarization.

"A line *h g* is also visible in the upper photograph which does not appear in the other.   If due to polarization, this confirms rather than contradicts the conclusions noted above.   It is not improbable, however, that this line may be due to a protuberance or bright spot in the corona, which is diffused in the ill-defined image.

"In repeating this observation, a Rochon prism of quartz should be used in which both images are equally corrected for color.   The angular separation need not exceed one degree, as it is only essential that the photographs should not overlap."

After Professor Pickering had completed his examination (in entire ignorance of the direction of the axes of the rhomb), I received from Dr. Lytton the prism with which the photograph had been taken.   Professor Pickering and myself then confirmed Dr. Lytton's statement that the plane of polarization of the achromatic image is nearly parallel to the line joining the two images, — a conclusion which points to the tangential polarization of the light of the corona.

Dr. C. S. Hastings has since courteously complied with my request, and made a critical examination of the photographic negative; and, though he confirms the description of the appearance of the negative, he dissents from the conclusion that there is positive evidence offered regarding the polarization of the corona.   It would be beyond our province to do any thing more in this report than simply to record the phenomena observed.   It is not possible to reproduce the photograph in question, and I can only leave the discussion to the physicists, with the remark that the negative, as described by Professor Pickering and its details corroborated by other competent witnesses, should be put in evidence as regards the polarization of the coronal light, unless the method can be shown to be open to serious objection.

Professor PICKERING, Dr. LYTTON, Dr. HASTINGS, Mr. WELLS, Mr. TOLLES, and Mr. EDWARDS have all rendered important service in the preparation of the above remarks relating to the photographs, which I desire to gratefully acknowledge.

                                                                LEONARD WALDO.

REPORT OF MR. R. W. WILLSON.

In accordance with the plan of the expedition, I proceeded to Fort Worth in advance of the other members of the party, arriving on July 9th. The first week after my arrival was spent in examining various points in the neighborhood of the town, in order to select a place for our observations. After the arrival of the " Brown Transit " and the chronometer, Bond, 1058, on the 15th, I began a series of observations for time, which was continued until the 31st.

In the division of labor upon the day of the eclipse, the duties that fell to me were of a general nature. I was to observe the times of the several contacts, and to examine the corona with the polariscope. It seemed best, also, that some one should watch the progress of totality, and obtain a telescopic view of the corona, examine its general structure, and take note of any new or remarkable phenomena which might present themselves. This duty naturally devolved upon me, not as being best fitted by reason of previous close familiarity with the appearance of the sun under all conditions, but as the only observer who had not already assumed some special object for his attention during the eclipse.

I was therefore provided with the three-inch telescope, by Secretan, with universal mounting, and a list of questions prepared by Mr. L. Trouvelot, of Cambridge, the answering of as many as possible of which was my principal aim.

For observation of the first contact, I made use of a terrestrial eye-piece, magnifying 42 diameters; and for the last contact, an inverting eye-piece, magnifying 98 diameters. Both powers as well as a power of 150 were used at different times for watching the progress of the partial phases. For second contact, and during totality, an inverting eye-piece magnifying 45 diameters was used, the diameter of the field of view being about sufficient to give a clear view of all parts of the corona lying within one diameter of the sun's centre. In this eye-piece was a glass plate ruled with concentric circles, about 6′ apart, whose circumferences were divided into arcs of 45° each, by lines radiating from their common centre, giving a simple means of locating the outlines of the corona and the positions of noteworthy portions. I was also provided with an Arago polariscope with selenite plate, and with a small direct vision spectroscope, by Browning, to be used if occasion should offer.

First contact was noted at $3^h 12^m 25^s$, and was a very good observation, the sun's limb being very clearly defined and steady, and the irregularities of the moon's limb very marked. Indeed, throughout the whole time of the eclipse, till a few minutes before the last contact, the atmospheric disturbances were remarkably small, and the definition such as left nothing to be desired.

As the moon advanced, a careful examination of the sun's surface showed no distortion of its details which could be attributed to a lunar atmosphere; the faculæ retaining their shape quite unaltered as they were covered by the advancing limb. The cusps were throughout sharply defined, no brushes of light or other similar phenomena were noticed; but, at times, the points appeared to be slightly curved outwards, an illusion which was less marked with a higher power.

At $4^h 16^m$ the extremity of the lower cusp was separated from the remainder of the visible portion of the sun. As totality was now fast approaching, I replaced my dark shade-glass by a lighter reddish one, in preference to a very light green shade, for which the light was still too strong. To avoid the necessity of another change, I determined to observe the second contact with this, perhaps, objectionable color; and therefore caught up the beat of my chronometer, and waited for totality, watching with my left eye, my right having been for some time protected by a bandage.

The progress of the moon now seemed very rapid; small portions of the cusps, now reduced to thin lines, being broken off by the projections of the moon's limb, and formed into little blotches of light evidently much increased in width by irradiation, which gradually lost their brightness without much change of size or shape, fading till they disappeared. This continued till only a thin strip of light remained, and this at last broke up into several fragments, which in the same manner faded gradually to extinction some four seconds later. The time of losing the last ray of light was, however, very well marked, and was noted as $4^h 18^m 20^s.6$.

Totality having commenced, I removed the bandage from my right eye and examined the corona. My first glance caused me great surprise. The moon appeared surrounded by a faint reddish glow, with a slight orange tinge, perfectly uniform in structure at all parts of the disc, brighter towards the inside, and diminishing in brightness rapidly and regularly outwards to a pretty well-defined limit about four or five minutes from the moon's limb. No prominences were visible. This was my first, and a very disappointing, impression, as I had been led to expect a variety of detail, which was entirely wanting.

Just as the time-keeper called 2.15, however, I noticed that I had forgotten to remove my shade-glass, which I now immediately did, regretting that I had lost 15 seconds of precious time.

Not altogether lost, perhaps; for, on removing the shade, I became aware of other portions of the corona, whose light was nearly as intense as that near the moon's limb, but which could not be seen through the shade; while the ring, which alone was visible through the shade, was not distinguished from the other parts of the corona. Moreover, as seen without the shade, the light adjacent to the moon was of a very different intensity at different parts of the ring.

This shade absorbs all rays beyond the line 1350 of Kirchhoff's scale; and a hasty examination shows that it transmits 17.6 per cent of those rays which correspond in refrangibility to the line a of the solar spectrum, 9.5 per cent of the B, 2.1 per cent of the C, and 0.4 per cent of the D rays.

On removing the shade, the moon appeared as a bluish-gray disc of uniform color and brightness, projected on the corona as a background. The light of the corona varied considerably in different parts, and its outlines were well defined, but very irregular, as will be seen by reference to Plate II., which represents its appearance as seen with an inverting telescope. At the point a, 180°, and at b, about 130° from the vertex, the appearance was as of cumulus cloud of rounded outline, and of pale rosy light, with very soft gradations of brightness, blending together at their point of meeting, and reaching about 13' from the moon's limb. The band *c* was nearly parallel to a tangent drawn at an angle of 315° from the vertex, and with its outside edge about 5' or 6' from the moon's limb at the point of its nearest approach. Both parts were brighter toward the moon's limb, giving somewhat the appearance of hollow cloud masses at *a* and *b*. The brightest points were at *a*, and at the western extremity of *c*, which were of nearly equal brilliancy. The light of these portions was perfectly continuous; and I found no trace of granulated or striated structure, and nothing whatever of the nature of streams or rays, though I devoted the greater part of my time to a close examination of these points, and particularly of *a*. At *d*, between 210° and 270° from the vertex, was a very much fainter light, with, perhaps, a tinge of green, conveying a very decided impression that it was of a totally different character from the other portions of the corona.

I regret that I had not time to examine the region more carefully. I followed the light certainly more than 45' from the sun's disc; but as it was very faint, and as I lost all standard of comparison when the moon and other portions of the corona were out of the field, I examined it no further, fearing to lose the disc entirely on the return, as neither the motion in altitude or azimuth was perfectly smooth and regular. At the point marked *f*, I merely glanced; the light was of the same character as at *d*, but I am of the impression that it nearly faded out while still within the field of the telescope.

Several prominences were visible, at the points indicated; one of them, about 275° from the vertex, being of considerable size and sickle-shaped. I should have passed them by without much notice, had I not been surprised by their color. I had expected them to appear very nearly as seen in the *C* line, by means of the spectroscope with a tangential slit; but they could certainly not, by any stretch of the imagination, be described as of a brilliant ruby color: they were rather of a pale pink. I was forcibly impressed, too, by the similarity in color between the prominences and the points *a* and *b* of the corona. The tint was identical, though the light of the prominences was very much more intense. The color of the corona was much less evident in the brighter portions near the sun's disc, helping to give the impression of a hollow dome at *a* and *b*, to which I have alluded. The large prominence, 270° from the vertex, was carefully located as a point, by means of which to determine the true position of the radial lines.

Having partially satisfied myself as to the form and general appearance of the corona, and not having heard the warning from the time-keeper of the close of totality, I now took up the polariscope, and turned it toward the sun; but I had hardly placed my eye at it when the sun reappeared, and I saw that I had lost the third contact. I glanced into the telescope to make sure, and then noted the time, $4^h 21^m 2^s$, probably several seconds late. My misjudgment of the amount of time remaining, and my failure to hear the signal of the time-keeper, prevented me from any verification of my observations of the corona, and a more careful examination of the portions *f* and *d*. I immediately consulted a scale of colors which had been prepared for comparison with the light of the corona, and fixed upon a tint which very nearly resembles it: it is number 33 of the samples of Scheiffele sheet wax prepared by G. H. Smithers.

Shortly before the end of the eclipse, light clouds covered the sun, not obscuring it, however, sufficiently to prevent observation of the last contact, which took place at $5^h 20^m 33^s$. Applying the clock corrections, I find the following times of contact : —

Fort Worth mean time.

	h.	m.	s.
I.	3	11	10.7
II.	4	17	6.1
III.	[4	19	47.]
IV.	5	19	18.3

My thanks are due to Mr. Charles Taylor of the "St. Louis Globe Democrat," who kindly acted as my assistant and recorder.

R. W. WILLSON.

REPORT OF PROFESSOR J. K. REES.

INTRODUCTION.

IN the arrangement of the work of the party, it was decided to send the instruments to be used by me to St. Louis first, so that I might become acquainted more thoroughly with the apparatus I was to employ before leaving for Fort Worth. On the afternoon of July 6th, a case of instruments arrived at Washington University. This case contained a Transit Instrument by Temple of Boston, also a fine two prism spectroscope by Grunow of Columbia College, New York City. The transit was set up by Mr. Willson and myself, on the transit pier of the observatory belonging to Washington University, and placed approximately in the meridian. On July 9th, other boxes of instruments arrived. These cases contained a five-inch refractor by Alvan Clark & Sons, with tripod stand. The spectroscope which had arrived previously was now adapted to the telescope in the manner hereafter to be described.

The telespectroscope was set up and adjusted. Considerable time was given to examining the comparative effects of the two sets of collimators and observing telescopes. Every opportunity was taken to become reacquainted with the exact positions of the Fraunhofer lines. In former years, I had spent much time in spectroscopic work, but I had not used a spectroscope attached to a large telescope.

On the night of July 16th, I left for Fort Worth, in company with Mr. Waldo and Mr. Seagrave. Through the generosity of Adams Express Co., I was able to send my boxes of instruments through to Fort Worth free of charge. This favor was all the more acceptable, because the boxes arrived the day we did, whereas freight packages were delayed a week. On our way down to Dallas, at the request of Mr. Waldo, I stopped off at McKinney. Here I left a stop watch and instructions with Mr. W. H. Chandler, who agreed to make observations on the duration of totality. This station, being near the limit of shadow, would give important results. Mr. Chandler interested others with him in the matter. Editors of papers, and the people generally, were ready and willing to do every thing in their power to assist the party in their work.

I took with me the following

INSTRUMENTS.

1. *A Fine Achromatic Telescope,* made by Alvan Clark & Sons of Cambridgeport, Mass., and described by Mr. Waldo as No. 1 of the list given in the Introduction.

The mounting was solid, but rather clumsy. The stand was of such construction as allowed only motions in altitude and azimuth; and the slow motion arrangements were of such a character as made it necessary to "set back" every hour. I had attached to this telescope, by means of iron collars, a board of pine, 1828.77 mm. long by 203.19 mm. wide, and 9.5 mm. thick. The board projected beyond the eye-end about 609.5 mm. On this projecting piece was adjusted the spectroscope. A covering of black lustreless silk was fastened closely over the prisms to keep out all light except that coming from the collimators. The slit of the spectroscope was uncovered, so that I could always look at the image of the sun, &c., on the slit plate. The finder of this telespectroscope was not used. I pointed the instrument by noting the position of the sun's image on the slit plate.

2. *Spectroscope*, made by Grunow of New York City, and loaned me by the owner Mr. Waldo.

The general arrangement of the instrument was after the usual pattern. The two prisms were, however, extra heavy and large. The slit plate, as was the case of the whole instrument, was black-lacquered, and images of the sun were remarkably distinct on the face of the plate. The edges of the slit were beautifully ground and very true. The slit could be varied in width by a screw made with fifty threads to 25 mm. The head of this screw was 25 mm. in diameter, and was provided with a scale; and an index showed the fractional part of a revolution, thus giving the width of the slit opening. The spectroscope is supplied with a double set of collimators and telescopes, also a collimator for a scale. The general action of the instrument is too well known to need explanation here. The measuring scale was of tin foil, through which fine lines had been cut, equally spaced, by a dividing instrument. This scale was cemented to a thin strip of glass, thus allowing the passage of light through the divisions in the tin foil, and forming at the focus of the observing telescope bright line images of the scale divisions. By adjusting the glass to which the scale was cemented, the scale divisions could be thrown above or below, or on the spectrum, to be examined. The glass carrier was also capable of lateral motion. A micrometer screw worked against the frame in which the scale was fitted, and by means of this screw the scale divisions could be made to traverse the whole field of the spectrum of sunlight, except a portion of the extreme violet end. To the frame that carried the scale there was attached an index which moved over a fixed scale, which was so graduated as to make one full revolution of screw-head equal to one of the divisions of this fixed scale. The screw-head was graduated also, and could be read to the hundredth of a single revolution.

Thus, for example, the scale-frame being set so as to bring the divisions pretty nearly to the middle of the spectrum, the position of the index was noted and recorded. Generally, I fixed the index exactly to the twelfth division of the fixed scale, the screw-head index marking zero. As I looked into the observing telescope, there was a division on or near the line to be measured, except for lines in the extreme red and violet. If a division was exactly on, or was a continuation of, the line to be measured, the number of that division counted from the red end was called off. If a scale division did not exactly coincide with the line to be measured, the screw-head was turned, so that the next division toward the red end of the spectrum was moved up to coincidence with the line. Then, calling out the number of the division, my assistant, Mr. Lomax, read the screw, which gave the fractional part of the space between two divisions that the measured line was beyond the scale division I had called off. When the position of the line to be measured was beyond either end of the scale, as seen in the observing telescope, the last or first division (as the case might be) was called off, and the screw-head turned till the last or the first division came into coincidence with the line to be measured.

Then the fixed scale over which the index of the frame moved was read, to see how many whole divisions had been passed over, and the screw-head was read for the fractional part of a division. This was the method used for exact measurement. Simply noting between what divisions the lines observed came, a close approximate position could be recorded. The former method I had decided to use in measuring the position of any line in the spectrum of the corona that appeared new to me; the latter method for the previously observed lines. The attachment of the spectroscope made it impossible for me to revolve the spectroscope on an axis, so that the slit could be placed tangential at two points, normal at two points, and between these two positions at all other points of the sun's image.

The constants of the Spectroscope are given in the list of instruments in the Introduction, No. 10.

Mr. Waldo loaned me for use with this spectroscope a battery of very fine eye-pieces by Crouch. I used during totality one of these eye-pieces, magnifying 11 diameters.

3. *Stackpole Sextant and Artificial Horizon.* This instrument I used in sun observations for latitude, and is No. 17 of the list in the Introduction.

4. *A Stop-watch,* to be used in noting times of contact.

WORK BEFORE THE ECLIPSE.

The days before the eclipse were occupied in the general arrangement of apparatus, and specially in the following ways:—

1. Careful adjustments of the telespectroscope.

2. Measurements of the positions of the principal Fraunhofer lines.

3. Practising with Mr. Lomax, who had kindly offered to assist me, in reading off the above measurements rapidly and accurately.

4. Sextant observations on the sun for latitude.

5. Acting as recorder for Mr. Willson in time observations with the transit instrument.

6. Comparing our time with Washington time by means of telegraphic signals sent from the Naval Observatory.

I shall speak of each point in detail.

1. Inasmuch as we had no building in which our instruments could be set, adjusted, and left standing, we were compelled to take down our apparatus every day, and often I was obliged to detach my spectroscope from the telescope, in order to place the instruments in the store-room kindly furnished by Mr. Lomax. Frequently, therefore, the adjustments had to be carefully examined.

2 and 3. Mr. Lomax and I practised, together and separately, reading the position scale. We measured the Fraunhofer lines, and also measured position of bright lines given by alcohol flame, to which were supplied various salts.

4. The sun observations for latitude were begun on July 20th, and continued to July 25th inclusive. It was found that, using the sextant stand devised by Mr. Willson, observations could be made on the stars with considerable increase in accuracy. Mr. Waldo made these latter observations, as at the time I was assisting Mr. Willson with the time observations.

The sun observations for latitude were conducted as follows: —

All observations for latitude were necessarily made about one hour and a half or more from the meridian. The declination of the sun being about + 20°, and the latitude of Fort Worth being + 32° 44', the sun at noon reached an altitude of 77° 16'. Its double altitude would be 154° 32'. As the sextant limb could be used only to 130°, the hour angle from the meridian was generally greater than 22° 30'. It was therefore impossible to use "circum-meridian altitudes" of the sun, and hence the method adopted was mainly that of equal altitudes. These observations can be used to determine the clock error and rate, though for that purpose it were better to observe the sun when moving faster in altitude and nearer the prime vertical. The transit observations make it unnecessary to reduce these observations for time. In some cases, the set corresponding to the morning observations was lost by reason of clouds or delay in getting to the place of observation. The manner of taking these observations for latitude was as follows: "Suppose the sun to be rising. The index of the sextant was set at an angle a little greater than the double altitude of the upper limb, the angle chosen being always at some whole degree, or else at some 10', 20', or 30' on the graduated scale. Then, looking through the telescope of the sextant into the artificial horizon, the two images of the sun were seen separated, but approaching each other, and the instant when the two limbs came into contact was noted. The index was then set forward 10', and the same process repeated, after which it was again set forward 10' more, and the same process repeated a third time, thus giving three observed contacts of the upper limbs. The roof of the artificial horizon was next reversed, and the index of the sextant set back to its first reading. Then, looking through the telescope of the sextant into the artificial horizon, the two images of the sun were seen overlapping, but separating, and the instant of last contact of the two limbs was noted. The index was then set forward 10', and the same process repeated; after which, it was again set forward 10' more, and the same process repeated a third time, thus giving three observed contacts of the lower limbs, respectively, at precisely the same altitudes as the corresponding ones of the upper limbs." Five and even six contacts of each limb were sometimes taken. In the afternoon, at the time when the sun reached the altitude it obtained in the latest morning observation, a similar set of observations was begun, only their order was reversed.

Mr. Willson or Mr. Waldo always noted the times at my signal.

The reduction of the observations for latitude has been effected by means of the following formulæ (using Chauvenet's notation): —

$$\text{Tan } D = \frac{\tan \delta}{\cos \frac{1}{2} \lambda}$$
$$\cos y = \frac{\sin h \sin D}{\sin \delta}$$
$$\varphi = D + y$$

$$\lambda = (T^1 - T) + (\Delta T^1 - \Delta T) - (e^1 - e)$$

Where T^1 = mean time of afternoon observations.

T = mean time of morning observations.

ΔT^1 and ΔT = corrections of T^1 and T.

e^1 and e equations of time at T^1 and T.

δ = declination of ☉ at the mean of T^1 and T.

h = altitude, corrected for refraction, &c.

The results of my latitude observations are given in the Introduction, p. 25.

5. Every fair evening I acted as recorder for Mr. Willson in his transit observations of stars for time.

6. The party had made arrangements, through the kindness of the Western Union Telegraph Company and the Texas Pacific Railroad Telegraph Company, to receive the Washington time signals. As the two lines had to be connected at Dallas, Texas, and Mr. D. P. Todd of the Nautical Almanac Office was to make observations at that place, he offered to see that the signals came through to Fort Worth. The line from Dallas to Fort Worth being used constantly by the railroad, it was impossible to prevent frequent interruptions of the signals. On July 27th, Mr. Waldo and myself were at the telegraph office in Fort Worth at the proper time, but, owing to some fault in the line, the signals did not come through. On the 28th, the signals were constantly interrupted, and no good comparison was taken. Mr. Todd, however, had received the signals, and shortly after exchanged signals with Mr. Waldo.

On July 29th, I made comparisons.

The manner of making the comparisons and their results are given in the Introduction, p. 19.

DAY OF THE ECLIPSE.

The morning promised badly. The sky was heavily overcast, and there seemed little prospect for a good day. However, all instruments were set out in their arranged places. I had decided to view the partial phases with the same five-inch telescope to which the spectroscope was attached. For this purpose, a diagonal eye-piece was used. This eye-piece was focused very nicely, and allowed to remain fixed. By drawing out the reflecting wedge of the eye-piece, the focusing was not destroyed, and the image of the sun was formed on the slit face of the spectroscope. After arranging my apparatus so as to be ready to measure the position of the Fraunhofer lines, for practice and comparison, should the clouds become less dense and give us a glimpse of the sun or a blue patch of sky, I compared the mean time chronometer Bond, 1058, with the sidereal chronometer Kullberg, 1178, and went to the telegraph office to get the Washington signals previously spoken of.

On my return, a table was arranged close by the telescope, at which sat Mr. Frank Doremus of the " Galveston News," who had consented to act as recorder for me. Mr. Lomax's position was at the scale collimator.

Just before the time calculated for first contact, the sun came out gloriously between the clouds. At once, I began to observe the sun with the five inch, using a yellow-green shade and a magnifying power of eighty diameters, watching for first contact. First contact was lost on account of my carelessness in looking on the wrong side of the sun's image. However, when I did get a view of the moon's limb on the sun, I started my stop-watch, and estimated that I was one minute late in so doing. Shortly after, I compared my watch with Kullberg, 1178, and Bond, 1058, then stopped the watch and reset the hands to 0ᵐ 0.0ˢ.

Between the first and second contacts, I examined the eclipsed sun with the above-mentioned eye-piece. No phenomena of distortion of the moon's limb were noted. The moon's limb was sharply defined, though there was a very perceptible undulation of that part of the limb nearest

the sun's centre.   This undulation was not communicated to the cusp points, which were
remarkably well defined and clean cut to their very point.   The serrated edge of the moon
could be made out very distinctly.   I was struck with the appearance of faint dots of light,
almost like faculæ, at end of each cusp.   These dots remained in view during all the partial
phase.   I called Mr. Waldo to look at them, and he saw them also.   They were not like
"brushes of light," but dots, say 3" to 5" in diameter.   I did not try to obtain a spectrum of
them.   To my eye the dots were always present distinctly, though faint.   The time of totality
approaching, I removed the reflecting wedge, and the image of the eclipsed sun was finely
focused on the slit.   The complete serration of the following cusp was beautifully seen on the
black face of the slit plate.   The positions of the Fraunhofer lines were again measured and
recorded.   A few moments before totality, I covered my eyes with a handkerchief, and as soon
as " Time " was called, indicating that totality had begun, I removed the bandage, intending to
devote seven seconds to a naked eye view of the corona, so as to note at what points it was
brightest, in order to direct the telescope to those points.   Instead of spending only seven
seconds of the precious time in this view, I must (Mr. Lomax tells me) have given some twenty
seconds to the view.   I was disappointed in the brightness of the corona, but judged only from
what I had read, and the impressions pictures had given me.   It was of a beautiful soft silver
gray color to my eyes, and was wing-shaped, the two wings being situated in the ecliptic, as
nearly as I could judge.   The preceding wing was indented at a distance that seemed equal to
some 40' from the moon's nearest limb.   The following wing extended to a greater distance
from the moon's limb, and seemed to be the continuation of the preceding wing.   The following wing was not indented, but rather projected in that part where the other wing was indented.
The light of the wings appeared made up of rays parallel to each other, and to the line
of the ecliptic.   This parallelism was very distinctly marked, and a fine sight.   I was called
from this enchanting view by Mr. Lomax ; and, adjusting the slit radially on the preceding limb
of the sun, I observed a continuous spectrum, with no lines whatever.   The slit now was one-
twentieth of a revolution of the screw-head open = .0254 mm.   The slit was then adjusted radially to the following limb of the sun.   Still a continuous spectrum, and no lines.   I expected
to see bright lines, and began to feel very nervous, thinking that something might be wrong
with me or my instrument.   I thought my eye, perhaps, had been made non-sensitive to faint
lines by the look at the corona and my work during the partial phase.   I now opened the slit
to one quarter of a revolution = .1270 mm., and brought the image of the eclipsed sun so that
the north limb was almost tangent to the slit.   I thought the bright lines might appear when
the slit was wider ; but I obtained a bright continuous spectrum again, and this time I noticed
that it was crossed by *dark* lines, among which C and D were plainly visible.   The appearance
was so different from what I expected to see, and, some seconds later the sunlight flooding in,
I was at the time almost inclined to think I had been mistaken.   But the record was called
to Mr. Doremus, " Continuous spectrum, dark lines C and D visible," some ten seconds before
the flood of light came.

After third contact, I put in the reflecting wedge, and again with the diagonal eye-piece
examined the partial phase.   The mottled appearance of the sun was very distinct.   The cusp
points had sharp outlines, but both points were blunt and slightly bent, seemingly in the direction of the sun's limb.

The undulations along the middle of the moon's limb were very distinct, but not communicated to the cusp points. The faculæ and mottled appearance of the sun, it seemed to me, did not come in contact with the moon's limb, but stopped short of it. The limb was plainly irregular. Toward the last of this partial phase, the clouds came over the sun, and I looked at the sun without any shade. I noted as nearly as I could the time of the last contact, but was not very sure about it on account of the clouds.

I was prepared to use a bull's eye lantern to throw light on the divisions of the slit screwhead; but it was not dark enough to require the use of any artificial light.

CONTACTS.

Contact	Reading of Stop Watch.	Reading of Bond, 1058.	Error of Bond on 1058.	Deduced Time.	Remarks.
	m. *s.*	*h.* *m.* *s.*	*m.* *s.*	*h.* *m.* *s.*	
I.	5 0.0	3 18 26.0	+ 1 14.7	3 12 11.3	Estimated 1ᵐ late.
II.	10 15.0	4 28 47.0	+ 1 15.09	4 17 16.9	
III.	Not taken.				
IV.	3 16.2	5 24 0.0	+ 1 15.30	5 19 28.5	Very cloudy.

The stop-watch was *started* at the beginning of each contact observed (except the first), and reset when comparison had been made between the stop-watch and Bond, 1058.

To my assistants, Mr. Lomax and Mr. Doremus, I am much indebted; especially to Mr. Lomax for giving to me so much of his time for practising on the measurements of the position of Fraunhofer's lines.

J. K. REES.

REPORT OF MR. W. H. PULSIFER.

My allotment in the scheme of work of the Fort Worth Eclipse Party consisted of the spectroscopic observation of the contacts, and the study of the spectrum of the corona, particularly the 1474 line. I left St. Louis at 9.30 P.M. on Tuesday, July 23d, and arrived at Fort Worth on Thursday the 25th.

I took with me the following apparatus: a four-inch refractor by A. Clark & Sons, a ten prism direct vision solar spectroscope by Browning, of London, and a single prism chemical spectroscope by A. Clark & Sons.

The Browning spectroscope has sufficient dispersive power to show solar prominences in full sunlight. The Clark spectroscope has a very fine prism, easily separating the D lines. It has a long slit, and was thought to be well adapted to the study of the coronal spectrum.

It was originally intended to use both spectroscopes in connection with the four-inch telescope; but experiments showed that, under favorable circumstances, the spectroscopes could not be changed in less than twenty seconds, thus entailing a loss of forty seconds of totality. To save this, on Saturday night, it was decided to attempt the adjustment of the Clark spectroscope to a three-inch telescope which had been brought to Fort Worth by Mr. F. E. Seagrave. By some unaccountable mischance, the tail-piece of this instrument had been left behind, and it was lying in our instrument room, a useless piece of apparatus, much to the regret of every member of the party. Arrangements were at once made with a turner to fit to this instrument a block of wood, bored to receive the support of the spectroscope. This was completed next morning, and I then adjusted to the telescope the finder of my four inch. This finder was provided with a needle point in the focus of the eye-piece, as described by Professor Harkness in his report of the 1869 eclipse.

After completing our adjustment, we contemplated our work with much satisfaction. It promised to add a valuable piece of apparatus to our battery. But the tripod stand belonging to the telescope proved, upon inspection, to be a very shaky affair. It had seen long and hard service, and had evidently suffered much from neglect. It was provided with "jerky" altitude and azimuth motions, was insecure and unsteady to the last degree, and ultimately proved a veritable snare and delusion. Still, with all its defects, it was considered better to risk it than the certain loss of forty seconds of totality and more, should an accident occur, or should the observer or his assistant be agitated when making the change. Thus arranged, my apparatus may be briefly described as follows: —

I. Telespectroscope, consisting of my four-inch Clark telescope and my Browning solar spectroscope.

II. Telespectroscope, combining Seagrave's three-inch telescope and my Clark spectroscope.

I was assisted by Mr. J. J. Roche and Mr. McHaddon, of Fort Worth. Mr. Roche assisted with telespectroscope I., and Mr. McHaddon had special charge of II. To these gentlemen I am under obligations for their faithful efforts. Unfortunately, they were assigned to me as assistants only the day before, and cloudy weather prevented preliminary practice.

At 2.30 P.M., July 29, I compared my pocket chronometer Jurgensen, 13548, with the sidereal chronometer Victor Kullberg, 1178, with the following results: —

	Victor Kullberg, 1178.				Jurgensen, 13548.		
	h.	m.	s.		h.	m.	s.
July 29ᵈ.08	12	47	0.0	=	2	33	57.6

[whence the error of J. 13548 at this date is — 0ᵐ 25′.4 — L. W.]

At three o'clock, both instruments being in position, I seated myself at telespectroscope I., having II. on my left, conveniently placed to seize as soon as totality had begun. Both instruments were carefully focussed; and the slit of I. was placed, as near as we could determine, upon that portion of the sun's limb where the moon would first make her appearance. Mr. Roche assisted in directing the instrument; and Mr. McHaddon, with chronometer in hand, was ready to note the time. With the C line beautifully defined, I waited for the contact. Presently I heard Mr. Seagrave call out, "Chromosphere gone!" and Mr. Willson remark, "Here she is!" My C line was still visible, and remained in view some ten seconds. I had lost the contact.

I now removed the spectroscope from I., adjusted a solar diagonal with a power of thirty, to observe with the telescope direct. The definition was simply perfect. The granulations were distinctly seen all over the sun, and many faculæ were observed.

The serrated edge of the moon was very sharply defined, and the lunar mountains were beautifully distinct. The cusps were as clearly cut as possible, and I noticed no brushes of light on or near them. The granulations and faculæ appeared well defined, close up to the limb of the moon.

After a half hour's observation with the telescope, the solar spectroscope was replaced, and observations were made for new prominence lines. None were visible.

The atmosphere in the direction of the sun now seemed perfect for spectroscopic work. The clouds which had covered the sky during the morning had broken, and immense cumuli were piled up in the south, the east, and north-east, while the region occupied by the sun was clear and remarkably transparent. I have always secured the best results under these circumstances, and I think I have never had better definition than on this occasion.

About three minutes before totality, the slit of I. was carefully placed tangential to that portion of the sun's image where second contact would occur, narrowed down, and carefully focussed until the lines came out perfectly sharp and clear. The field covered the C line, jammed well over to the left, and ran up to W. L. 6300 or thereabouts. Mr. Roche, having a firm hold of the instrument, directed his whole attention to retaining the slit in position; while, seated in a chair, with my elbows properly braced and my eye at the eye-piece, I assisted in keeping the limb of the sun on the slit, and waited for the second contact.

The last moments seemed interminable. The excitement was intense. The strain upon my eye was so great that I feared my sight would fail me entirely before the contact. At last there came a flash, and the lines blazed out brightly, appearing to fill the field completely for an instant. Although expecting this phenomenon, I was fairly surprised; and I was startled, also, to find the bright lines *shortened at each end*, and occupying but about one third of the width of the spectrum. I was totally unprepared for this, but was instantly reminded of Miller's maps of the ultra violet spectrum of platinum as figured in Lockyer's work on Spectrum Analysis; but the lines were sharp and well defined, very narrow, and not at all nebulous, except that the ends seemed rounded off. *The C line did not shorten*, and was in view after the other lines had disappeared.

The reversal of the lines was so sudden and evanescent, and the shortening so astonishing, that I utterly failed to notice the length of time they were in view; nor did I observe any continuous spectrum, unless the first flash, which seemed to fill the field with light, might be taken as such. I am not certain that all the lines were reversed.

Upon the appearance of the reversal, I shouted to Mr. Roche, who recorded the time, 4^h 16^m $49.^s6$.* This may be a second or more late, as it was quite impossible to recover at once from the surprise excited by the phenomenon.

I immediately seized an opera glass to view the eclipsed sun, that I might direct Mr. McHaddon where to point telespectroscope II. I had allowed myself twenty seconds for this; and at the expiration of this time, having changed my place and being seated at the eye-piece of II., I instructed my assistant to direct the instrument to a portion of the corona which seemed promising. But here I have nothing to record but disappointment and failure. At 4 P.M., as previously arranged, the lantern which was provided to illuminate the scale of the Clark spectroscope was lighted and seemed to burn well, but just before totality it was discovered that it was extinguished. My assistant was obliged, therefore, to leave his instrument, and seek other means of illumination. He did not return until after the last rays of the sun were obscured, and the image on the slit was lost.

Here the deficiencies of the mounting made themselves felt. Mr. McHaddon endeavored to bring the point I had shown him upon the slit; but the unsteadiness of the mounting, added to the nervousness of an inexperienced operator, utterly prevented it, and a minute passed in fruitless efforts to properly point the instrument. At last, and just as in desperation I had risen to direct it myself, Mr. Roche came forward and expressed his ability to operate it. Believing that his experience and success with the other instrument would enable him to assist me here, I again seated myself at the eye-piece. The second attempt was no more successful than the first; and the sun burst forth, finding us still struggling with our imperfect apparatus.

Shortly after the sun reappeared, light clouds floated over it; and, at the fourth contact, the absorption was so great that the chromospheric lines were invisible.

I was very much puzzled by the shortening of the reversed lines. I had met with no account of such a phenomenon, and my first thought was that some unaccountable derangement of my apparatus had caused it. After third contact, I carefully examined my instruments, but found every thing in perfect order, and I confess I knew not where to look for a disturbance which would produce such a result. At last, it dawned upon me that the only rational explanation would be found in the fact that the image of the sun formed on the slit was so small ($0^{in}.54$) and

---

* Which would be 4^h 17^m $15^s.3$ Fort Worth mean time, assuming the correction of J. 13548 for July $29^d.16$ to be $+ 0^m$ $25.^s7$. — L. W.

the reversion layer was so thin that *my slit extended beyond the layer on each side*. This, and this only, it seemed to me, would fairly explain both the normal condition of the C line and the shortening of the others. It then occurred to me that this combination of circumstances would give me a method of measuring the thickness of the reversion layer as it was represented by the lines in view.

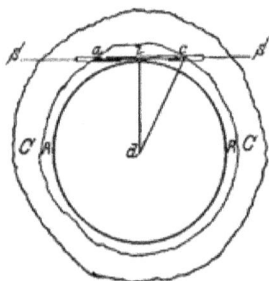

In the accompanying diagram, the proportions of which are of course exaggerated, *C* represents the chromospheric layer surrounding the sun, and *R* the reversion envelope; *S*, the slit of the spectroscope; *a, b, c*, the shortened lines; and *d, b*, the radius of the sun. It is evident at a glance that the thickness of the reversion layer must be equal to the difference in the length of the lines *d, b*, and *d, c*. In my telespectroscope, the diameter of the sun's image on the slit was $0^s.54$, and the total length of the slit $0^s.08$. The shortened lines occupied ⅓ the width of the spectrum. Now, assuming the sun's semi-diameter to be 430,000 miles, it is easily determined that the difference in the length of the lines *d, b*, and *d, c*, is 524 miles, which represents the minimum thickness of the reversion layer. I say minimum thickness, as the known lines in the field were principally those of iron and the heavier elements; and although I had the slit just touching the limb of the sun on the instant before the reversal, and am confident my observation was a good one, still I cannot be absolutely certain of the fact.

My spectroscopic work prevented me from looking for the corona before totality. I observed the eclipse of 1869 from my observatory in St. Louis, some fifteen miles south-west of the line of totality, and saw the corona and several prominences. A sketch made at the time has been mislaid or destroyed; but I append an extract from an article furnished by me to the "St. Louis Democrat," and printed in its issue of August 10, 1869.

St. Louis, August, 1878.   W. H. PULSIFER.

. . . Just at the moment of greatest obscuration by the moon, light clouds, that had for a few moments partially hidden the sun, passed away, and the corona seemed to *burst out* from around the moon, showing its spherical shape so perfectly that it seemed like a huge ball; . . . while above it a brilliant line of sunlight blazed out so narrow that I was enabled to remove the colored glass from before the eye-piece, and observed without experiencing any unpleasant effects to the eye. Scarcely had the corona appeared, and before any calculation had been made of its extent, there seemed to leap from the edge of the moon, and in close proximity to each end of the line of sunlight, large prominences of a beautiful rose color. I saw two near the eastern, and two near the western ends of the line of sunlight. . . .

These observations were made with a four-inch Clark Telescope. . . . *St. Louis Democrat*, Aug. 10, 1869.

REPORT OF MR. F. E. SEAGRAVE.

I LEFT Providence in company with Mr. Waldo of Harvard College Observatory on the morning of July 13th. We arrived in St. Louis on the morning of July 15th, where we spent two days, principally for the purpose of exchanging longitude signals between there (St. Louis) and Fort Worth. I left St. Louis on the evening of July 16th, and arrived in Fort Worth on the afternoon of July 18th. A site for our temporary observatory was immediately selected on the grounds of S. W. Lomax, Esq., situated about one mile south of the town. After my arrival at Fort Worth, I was occupied in mounting my instruments, taking observations of the sun, and making the meteorological observations which were taken three times a day. I was provided with the following instruments : —

1. A TELESCOPE.

The focal length of this instrument is 81 inches, the diameter of the object glass is 5 inches. The eye-piece used in nearly all of my observations has a magnifying power of 61 diameters, and is fitted with two shades of neutral tint glass. There are two other eye-pieces which have powers of 100 and 134 diameters. This instrument is mounted equatorially on a common tripod stand.

2. A SPECTROSCOPE.

This instrument has a system of four whole and two half prisms of 60° flint glass, which gives a dispersive power of ten prisms, by reflection. It was made by Browning of London, and is the instrument I used in all of my observations. It is provided with six eye-pieces, three positive and three negative. To one of the positive eye-pieces there was adapted an eye-piece micrometer, ruled by Professor W. A. Rogers, of Cambridge.

3. A SIDEREAL CHRONOMETER. (Victor Kullberg, 1178.)

At sunrise on the morning of July 29th, the whole sky was completely overcast. The clouds were in some places much less dense than in others, and the wind blew steadily from the south-east. At about ten A. M., the clouds began to break up and rapidly disappear. The sky was now clear and the temperature warm. About two P. M., large cumuli clouds began to rise in the south-west, and kept passing over the sun until about fifteen minutes before the time of first contact. The first contact was to take place at three hours and eleven minutes. At three o'clock, I took my place at my telescope, and pointed it at the sun. My observations of the first contact were made by placing the slit of the spectroscope tangential to the sun's limb, so as to watch for the disappearance of the C line in the spectrum of the chromosphere. At about three minutes before the computed time of first contact, I placed my eye at the eye-piece of the spectroscope, and brought the C line into the field of view. I watched it several minutes, and at 13h. 23m. 51.5s. by the chronometer the line began to grow shorter, and then went out, showing that the moon's limb had already occulted the chromosphere.* This disappearance did not occupy more than 2s., but it may have been much less. In a few seconds, Mr. Willson, who was observing next to me, observed the first contact with the sun's limb. After this, I was busy recording the temperature, which was taken every ten minutes, and in examining the sun's disc with the spectroscope. About 4h. mean time, which was about 16 minutes before totality, the clouds which were seen in the north and west began to disperse, with quite a change of temperature. The sky around the sun was very black, and every thing began to have a more gloomy aspect. I now made a very careful search for new prominence lines around that portion of the sun's disc which was unobscured, but saw none. About 4h. 17m. mean time, the last ray of sunlight disappeared. I now looked very carefully for the famous 1474 line of the coronal spectrum. At first, I saw nothing, but on widening the slit I saw one bright line which was quite faint and not very well defined: this was the only line in the field. Its measured width at its middle point was 1.3 divisions of the eye-piece micrometer. As I was unable to secure any other measures of this line before the end of totality, I have not determined the value of one micrometer division. I had hardly time to secure this measure of its width, and to record the temperature, when the sun reappeared. During totality, the telescope was pointed to the north-western portion of the sun's disc, position angle about 325°, and on the inner corona, or the portion of the corona nearest the moon's disc, which was much brighter than the outer corona. The darkness was not so great as I had anticipated, although a lantern had to be used to record by. Venus was shining brightly in the north-west, and the whole sky in the north and east had a most beautiful orange color. The thermometer was now at 81.5° F., a fall of 11.7 degrees since the time of first contact. Between the time of third and fourth contacts, I took the spectroscope off of the telescope, and applied a diagonal eye-piece, for the purpose of watching the reappearance of the faculæ and small spots, to determine whether they were as well defined near the moon's limb as they were some distance from it, for the purpose of detecting a lunar atmosphere. To me they appeared just as well defined. I then applied the spectroscope again to observe the last contact, which was to take place at 5h. 19m. mean time, but clouds covered the sun which were dense enough to prevent any spectroscopic observations.

The following are the thermometer readings, in the shade, of Mr. Waldo's thermometer Casella, 29563: —

Fort Worth, m. t. A. m.	Thermometer.		Fort Worth, m. t. A. m.	Thermometer.	
3 11	93°.2	First contact.	4 21	80°.5	
3 21	92°.2		4 31	86°.5	
3 31	92°.5		4 41	86°.7	
3 41	91°.7		4 51	87°.7	
3 51	90°.2		5 1	87°.7	
4 1	89°.0		5 11	88°.0	
4 11	82°.2		5 19	88°.7	Last contact.
4 15	81°.5	Totality.			

* *Note by L. W.* — If we assume the error of V. Kullberg, 1178, to be + 1h 43m 4s.9, we have for the Fort Worth local times of contact, 11h 40m 10s.6 s. t. = 3h 10m 44s.4 m. t.

From July 23d until the day of the eclipse, I examined the sun from time to time, as it proved convenient to do so. There was no spot visible during this interval. A group of faculæ on the sun's western limb I first saw July 23d, but they had disappeared on the 25th.

The following meteorological observations were made with Mr. Waldo's aneroid barometer, and with a thermometer which was found to be sensibly coincident with Mr. Waldo's thermometer, Casella, 29563.

The index correction of the aneroid, which has been applied to all of the following observations, was determined as follows:—

By Mr. Willson at St. Louis, July 2d, to be $-$ 0.23,

By Mr. Waldo at Cambridge, August 4th, to be $-$ 0.21,

whence the mean index error $-$ 0.22 has been adopted.

The observations up to July 19th were made by Mr. Willson, the remainder by myself. A storm which began in the afternoon of July 28th prevented my going out to our observatory, and the record is somewhat incomplete on that account.

Date, 1878.	Barometer.	Therm.	Wind.	Form.	Clouds.	Amount.	Pos.
July 10 3 P.M.	.29.38	92.0			Cum.	.05	
11 10 A.M.	29.41	87.8			Clear.		
3 P.M.	29.32	91.7			Cum.	.2	
12 3 P.M.	29.37	91.2			Cum.	.2	
13 11 A.M.	29.43	92.0			Cum. Rain.		
14 11 A.M.	29.39	86.0					
15 10 A.M.	29.28	87.7			Cum.		
2 P.M.	29.21	92.0			Cum.	.4	
16 10 A.M.	29.20	88.3			Clear.		
3 P.M.	29.17	93.6			Clear.		
17 8 A.M.	29.25	85.3			Clear.		
3 P.M.	29.19	94.0			Light Clouds.		
18 9 A.M.	29.26	89.0			Clear.		
3 P.M.	29.44	94.0					
19 9 A.M.	29.32	93.0	W.	0	Cum.	.1	Horizon.
2 P.M.	29.34	97.0	E.	2	Cum.	.5	N.
8 P.M.	29.28	80.0	S.	1	Strat.	.3	S. E.
20 9 A.M.	29.43	89.0	S. E.	1	Cum.	.5	Patches.
2 P.M.	29.35	94.0	S. E.	1	Cum.	.5	Patches.
8 P.M.	29.34	85.0	N. E.	0	Strat.	.2	E.
21 9 A.M.	29.43	94.0	E.	2	Cum.	.2	S. & W.
2 P.M.							
8 P.M.	29.33	85.0	E.	1	0	.0	
22 9 A.M.	29.34	93.0	N.	1	Cirr.	.1	N. E.
2 P.M.	29.23	98.5	E.	2	Cum.	.1	Horizon.
8 P.M.	29.28	83.3	E.	3	Haze.	.7	N. E.
23 9 A.M.	29.34	89.0	E.	3	Cum. & Cirr.	.3	S. & E.
2 P.M.	29.24	94.3	E.	3	Cum. & Cirr.	.5	E.
8 P.M.	29.24	81.0	N. E.	1	Strat.	.2	N. & E.
24 9 A.M.	29.33	93.0	S. E.	2	Cum.	.3	N. & E.
2 P.M.	29.23	103.0	E.	1	Cum.		
8 P.M.	29.12	83.0	E.	1	Strat. & Cirr.	.5	N. & E.

Date, 1878.			Barometer.	Therm.	Wind.	Force.	Clouds.	Amount.	Pos.
July 25	9	A.M.	29.24	91.0	S.	1	Cirr.	.5	N. & W.
	2	P.M.							
	8	P.M.	29.00	82.0	E.		Strat.	.5	N. & W.
26	9	A.M.	29.08	94.5	S. W.		Cum. & Cirr.	.7	
	2	P.M.							
	8	P.M.	29.04	81.0	E.	1	Cum. & Strat.	.7	
27	9	A.M.	29.18	85.8	S. E.	1	Cum. & Cirr.	.5	E.
	2	P.M.							
	8	P.M.	29.24	86.0	N. E.	1	Cum.	.8	
28	9	A.M.	29.23	88.5	S. E.	2	Cum. & Cirr.	.8	

I left Fort Worth with Mr. Waldo on the evening of July 29th, and proceeded with him as far as St. Louis.

I omitted to mention that I saw the inner corona form on the n. p. side of the sun about 30' before totality.

F. E. SEAGRAVE.

Reports from other Observers

I.

East Tennessee University,
Knoxville, August 5th, 1878.

Mr. Leonard Waldo, Harvard University, Cambridge, Mass.

Dear Sir, — I have the honor to make the following report of the humble part I took in the recent observations of the Solar Eclipse, conducted under your direction at Fort Worth, Texas.

I was making a tour through the State of Texas, and had to pass on my return homewards very near the point which you had selected as a suitable locality for the conduct of your observations. The public journals had informed me of the existence of your party, and of the extensive preparations you were making at Fort Worth for a series of elaborate scientific observations. Curiosity, not idle however, impelled me to take advantage of so rare an opportunity to observe both a total eclipse of the sun and a series of skilfully conducted observations at the same time. This accounts for my being at Fort Worth on the 27th day of July.

On introducing myself to you and the gentlemen with you, I was much gratified at not being considered as an idle interloper ; and still more pleased when you informed me that I might be useful to your party.

At your request, I undertook to make a free-hand sketch of the eclipse at totality, as seen by the naked eye.

I made preparations for this duty, by drawing a circle two inches in diameter on a piece of ordinary sketching paper. The circle was blackened with a dark soft pencil, and intersected by vertical and horizontal diameters, which were prolonged about two inches beyond the circumference.

In order that my eyes might be sensitive, I avoided looking much at the eclipse during its progress ; and, in accordance with your instructions, I closed and covered my eyes for about ten minutes before totality. Just before totality, I took a look at the sun through a stained glass to see if I could detect a corona, but failed to discover any. From that time until totality, I kept my eyes open, but cast down to the ground, and shaded by my black felt hat.

While thus looking down, I noticed the rapid diminution of light and increasing indistinctness of shadows which heralded the approach of totality. Though prepared by these phenomena for the final obscuration of the sun, yet the *actual darkness* which came over the scene and the *complete* and *sudden disappearance* of all shadows were really startling, when the exact moment of total eclipse occurred.

Simultaneously with the above noticed phenomena, Mr. Britton called out, "Total." I cast my eyes immediately to the sun and began to draw. The configuration of the external corona was so simple and striking it required but a few seconds to sketch it. I then directed my attention to the inner corona, and made a tracing of its outline around the black disk of the moon. I then sought for striæ, streaks, protuberances, &c., with but little to record. I was just beginning to fix my attention on the subject of *color*, when *suddenly* the totality ceased and the glare of the sun was too strong for the naked eye. An impression was made upon my mind of a reddish glare in close proximity to the upper dark limb of the moon, but I could not define either the form or extent of this red color.

The prevailing color of the entire phenomenon was bluish.

The moon's disk was a deep "Payne's Gray" (a neutral tint compounded of indigo, Indian red, and black). The inner corona was a pure bluish-white, like the light of the star Sirius. The outer corona was a fainter, more bluish-white, and compared in luminous intensity with the inner corona about as one to three.

The color of the sky was a dark indigo blue.

To my eye there was a large area of luminosity extending from the extreme brightness of the inner corona to the extreme darkness of the surrounding sky, by a gradation too insensible to be indicated in a drawing. Within this luminous space hung the eclipsed sun, with its face covered by the nearly black moon and its clearly marked inner and outer coronas.

The radiant effect of the outer corona was very much like the rays of light and shadow cast by the sun when obscured by a dark cloud ; the inner corona resembling somewhat the silver lining of the cloud, but not so sharply defined.

The curious facts, that the radiating rays of the outer corona extended in only two directions, and were so slightly divergent as to be nearly parallel, I think, deserve especial notice.

Having determined to make for you as correct and complete a drawing as possible, I bent all of my energies to that one purpose, and made no observations not immediately and directly connected with the object I had in view.

All of which is respectfully submitted by

Your obedient servant,

S. H. LOCKETT,
Prof. Math., &c., East Tenn. Univ.

II.

[*The marginal numbers refer to the printed instructions mentioned in the Introduction.*]

1. (Name of Station.) Allen, Collin County, Texas.
2. (Date.) July 29th, 1878, Monday P.M.
3. (Bearings.) About one hundred yards north-west of centre of Texas Central Railroad Depot in Allen. About eight and one-half miles south-south-west of McKinney, the County seat, and twenty-four miles north of Dallas.
4. (Description of watch.) Silver watch manufactured by the American Watch Company, Waltham, Mass., Bartlett movement. Numbered 314429, seconds' hand attached.
5. (Washington noon-time.) Applied each day for Washington noon-time at Western Union Telegraph Office here, but failed to get Washington time. Learned from operator that McKinney obtained Washington noon-time, and the difference between that and Houston (Texas) railroad time was on the 29th, 1*h.* and 21*m.* So we obtained Houston railroad time as sent here at 12 M. July 29th. The times at which the contacts were taken is in Houston (Texas) time.
6. (Contacts.) 1st contact 3*h.* 19*m.* 8*s.*
2d	,,	4	23	54
3d	,,	4	25	39
4th	,,	5	20	54

   } Houston (Texas) time.

7. (Estimate of uncertainty.) The sky was partly cloudy, and at times (principally at 4th contact) thin transparent clouds passed over the sun, and possibly may have delayed one or two seconds in the calling of the fourth contact. The original record is forwarded with this, and contains all the figures, &c., as made at the time without any changes whatever. In the first contact, you will see that the seconds were first put down 12, then marked over with 10, then cancelled and 8 written above. The 8 is the correct figure. The first contact was seen by the observer through the smoked glass four seconds before called.
8. (Remarks.) The seconds and minutes were counted and recorded with exceedingly great care, and in a way in which it was almost impossible to be mistaken. The duration of totality was 1*m.* 45*s.*
9. (Signatures of Observers.)

H. A. HILL, *Observer with Smoked Glass.*
J. M. HOBSON, *Time Observer.*
L. M. RUSH, *Naked-Eye Observer.*

59

III.

[*The marginal numbers refer to the inquiries in the printed instructions.*]

1. Dallas City, Dallas County, Texas.
2. July 29th, 1878.
3. No. 1116 Jacksont Street, three quarters of a mile east of Court House.
4. Good Patent Lever Watch, set three hours before beginning of eclipse with chronometer of J. M. Oram, Jeweller of Dallas, which marked the correct Dallas time, as verified by Mr. D. P. Todd, with the Naval Observatory Instruments; eighteen-inch telescope, belonging to a Gurley Level; and smoked glass for naked eye.
5. Time explained above.
6. 1st contact 3h. 11m. os.

   2d   ,,   4 19 58
   3d   ,,   4 21 56
   4th ,,   5 20   0

7. There may be some inaccuracy as to the time of the first and fourth contacts, as it was difficult to determine the exact moments of contact, especially as regards the fourth; portions of the face of the sun being at that time slightly dimmed by fleecy filaments of cloud.
8. The inner *corona* was very bright, and its outlines distinct and well defined. Its structure was filamentous, the filaments being straight and perpendicular to the contour of the moon. The extent of the inner *corona* beyond the moon was one-fifth of the moon's diameter. The outer corona was about one-half the width of the inner, and its outlines not very distinct to the naked eye. I enclose a rough sketch of the corona.

JAMES GILES,   } *Observers.*
J. H. BREEZE, }

IV.

McKINNEY, COLLIN COUNTY, TEXAS,
July 30th, 1878.

MR. LEONARD WALDO, Cambridge, Mass.

DEAR SIR,— The following naked-eye and smoked-glass observations, made by the undersigned, on the occasion of the recent solar eclipse, are respectfully submitted for your consideration.

Our place of observation was a window in the second story of a house about a quarter of a mile north of the Court House in the town of McKinney, County of Collin, and State of Texas. McKinney is situated about thirty miles north of Dallas, Texas. A more exact description of the location we are not able to make, as we have not the proper and necessary surveys before us.

The instruments of observation consisted of a piece of smoked glass and an ordinary chronometer with minute and second hands.

The first contact or beginning of the eclipse, according to our time, was ten minutes and thirty seconds after three o'clock in the afternoon, which, as near as we could ascertain from the telegraph, was an hour and twenty-five minutes earlier than Washington time. It is probable the first contact commenced somewhat earlier, as it came a little earlier than we expected according to newspaper reports.

The second contact or beginning of totality commenced at thirteen minutes and thirty-two seconds after four o'clock by same time, and lasted one minute and twenty-seven seconds, or possibly a shade longer, as a little confusion occurred in the room at the time, owing to a difference between the two observers as to the exact time.

The fourth contact or end of the eclipse was fourteen minutes and thirty seconds after five o'clock.

During the eclipse, it was obscured more or less by fleecy clouds of a transparent nature.

By the naked eye, all three of the observers saw very distinctly a star, which we took to be Venus, from the map of the eclipse we have. One of the observers noticed another star somewhat obscure, which he took to be either Castor or Pollux, according to the map.

The corona had rather a silvery appearance during the totality, and "the glowing points of light hanging upon the edge of the black moon and glistening like rubies" were not observable, as we expected; but probably they were more or less obscured by the fleecy clouds, as before suggested.

We did not expect the eclipse to occur quite so early, and we expected a longer totality. According to newspaper reports which misled us somewhat, the beginning was to occur at 3*h*. 11*m*. Totality to begin at 4*h*. 14*m*., duration 2*m*. and 31*s*., and eclipse to close at 5*h*. 19*m*.

In closing, we would state, as our report will doubtless disclose, that none of us are professional or amateur astronomers.

Respectfully submitted,

J. M. PEARSON.
JOHN S. MOORE.
W. J. FINCH.

V.

MCKINNEY, TEXAS, July 31, 1878.

MR. L. WALDO, Cambridge, Mass.

DEAR SIR, — In compliance with a request of Mr. Rees of your party, I send the record of the time of the eclipse at McKinney, which I carefully recorded with the aid of the watch left with me. I was able to ascertain correctly the time of totality.

1st contact 3*h*. 8*m*. P.M.
2d contact 4*h*. 14*m*.
From 1st contact to totality, 1*h*. 6*m*.
Duration of totality, 1*m*. 29½ *s*.
Time at expiration of totality, 4*h*. 15*m*. 30½ *s*.
Time at close of eclipse, 5*h*. 14*m*. 30*s*.
Time from expiration of totality to close of eclipse, 59*m*.
Difference of time from 1st contact to totality, and from totality to close of eclipse, 7*m*.
Duration of entire eclipse, 2*h*. 6*m*. 30*s*.

The above is as near as I could possibly get by my time (*Sun Time*), and several other parties which I employed made almost the same as I did, but from what I can learn I got the most accurate record of any. We had a splendid view of the sun. I do not know the number of Planets seen. Some saw two, and some three; but the names I do not know, though I saw none of them, as I was busily watching the sun to get the correct time of totality.

Very respectfully,

W. H. CHANDLER.

VI.

★ TEXAS EXPRESS COMPANY, EXPRESS FORWARDERS.
BREMOND, July 29, 1878.

MR. LEONARD WALDO, Cambridge, Mass.

DEAR SIR, — There was a cloud covering the sun at totality, but we were able to record the figures given by a small spot of sunshine through the cloud distant from our position three hundred yards. The cloud being stationary, we started the timer at its extinction, and stopped at its reappearance.

W. QUARLES.

[*The original memorandum accompanying the above note gives for duration of totality 1m. 56½s. — L. W.*]

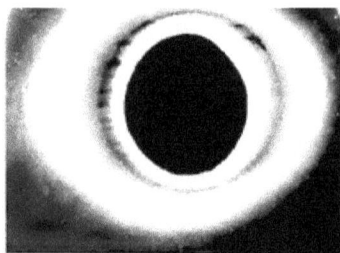

PLATE I.

Photograph No. 3, enlarged 3.2 diameters.

PLATE II.

To illustrate the Report of Mr. R. W. Willson.

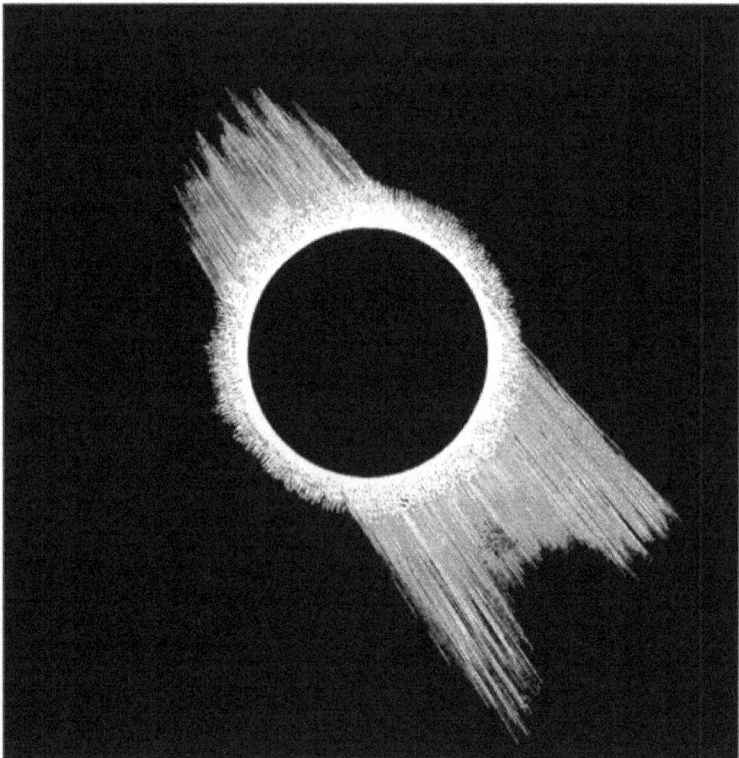

PLATE III.

Naked-eye View of the Corona.   Drawn by Professor S. H. Lockett.

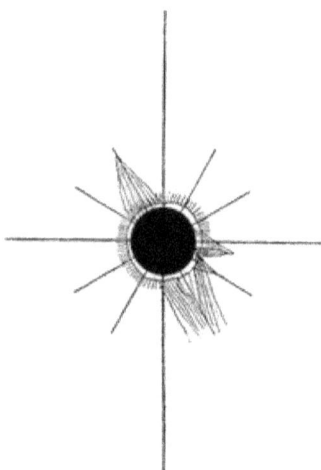

FIG. 1. — THE ECLIPSE, AS DRAWN BY MR. C. P. LEVY,
BUILDER, FORT WORTH, TEXAS.

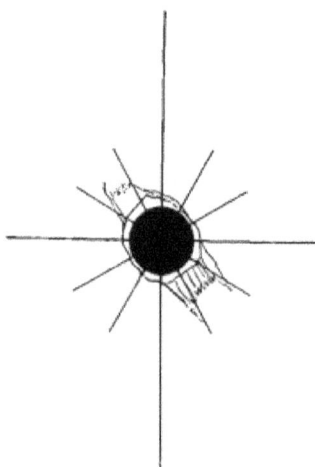

FIG. 2. — THE ECLIPSE, AS DRAWN BY MR. Z. H. POSTLES,
FARMER, FORT WORTH, TEXAS.

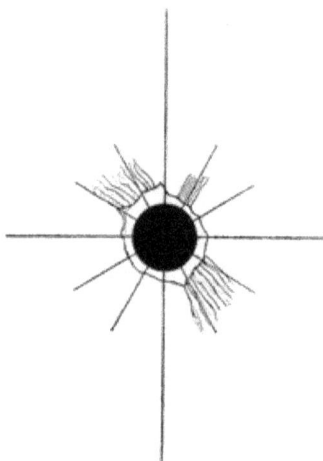

FIG. 3. — THE ECLIPSE, AS DRAWN BY MR. J. HAYS,
LAND LOCATOR, FORT WORTH, TEXAS.

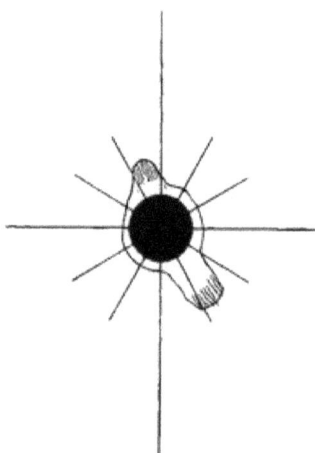

FIG. 4. — THE ECLIPSE, AS DRAWN BY MR. J. C. DENNY,
LUMBER CLERK, FORT WORTH, TEXAS.

PLATE IV.

www.ingramcontent.com/pod-product-compliance
Lightning Source LLC
Chambersburg PA
CBHW020337090426
42735CB00009B/1569